"Biodiversity"—Think About It!

1. In what way is the classification of organisms the same as the classification of habitats?

   _____
   _____
   _____

2. The largest group of habitats is biomes. There are many types of biomes. Examples include deserts, forests, grasslands, tundra, and oceans. Choose a biome and create a brochure using the science brochure planner. Make sure your brochure has the following information on the biome: worldwide locations, climate, plants and animals, an example of a food chain, and interesting facts.

3. Canada has many forests. One type is the temperate deciduous forest. Most trees in these forests lose their leaves in winter. These forests experience four seasons, with cold winters and warm summers. Deciduous forests receive precipitation all year and have fertile soil. Think about what you know about forests like this. Do research to find out more. On a separate piece of paper, draw then label plants and animals you might find in a temperate deciduous forest. Include some non-living things in your drawing.

4. How is the community you live in similar to a habitat community?

   _____
   _____
   _____
   _____
   _____

5. What do you think is the most important benefit humans get from biodiversity? Why?

   _____
   _____
   _____
   _____
   _____
   _____

Species Interactions

All organisms in a community have specific roles to play. Each species interacts with one or more other species. Most interactions involve competition for resources. Resources are things an organism needs to survive. These include food, water, sunlight, and shelter.

In a healthy community, there is a balance of organisms so each one gets what it needs. An example is the food pyramid. There are more species and individuals at the bottom of the pyramid. There are less and less as you move up the pyramid. When a community works the way it should, there is a balance in the number of species at all levels.

example of a food pyramid

Symbiosis

Some organisms have a special relationship with other organisms. Symbiosis is a close relationship between two or more organisms. At least one organism receives a benefit from this relationship. There are three categories of symbiosis.

- **Mutualism**—Both organisms benefit from the relationship. One example is between a zebra and a bird called an oxpecker. The oxpecker eats ticks and other parasites that live on the zebra's skin. The oxpecker gets food, and the zebra benefits from having the pests removed.

- **Commensalism**—One organism benefits from the relationship and the other is not affected. One example is the remora. This fish has an adhesive disk on its head. The remora uses this to stick to a larger animal such as a shark. When the shark is eating, scraps of food float away from the shark's mouth. The remora eats the scraps of food. The shark is not affected by the presence of the fish.

- **Parasitism**—One organism benefits from the relationship and the other is harmed in the process. One example is the flea. Fleas live on a host, such as a dog or cat. They bite their host's skin and suck their blood. The fleas get food and a warm home. But the host gets bitten repeatedly and is very itchy and uncomfortable.

"Species Interactions"—Think About It!

1. List these organisms in the order they would appear on a food pyramid, from bottom to top: mouse, corn, hawk, snake. Why would they be in this order?

_____

_____

_____

2. Read the following descriptions. Tell whether each gives an example of mutualism, commensalism, or parasitism. Give reasons to support your answer.

   a) Burdock is a weed. The seed heads of the weed are burrs with hooked tips. These tips catch on cows, deer, or even people as they pass by. The burrs are dropped or brushed off in different places.

   _____

   _____

   b) A certain type of bacteria lives in human intestines. This bacteria eats food that humans cannot digest, and partially digests the food. Then the human finishes digesting the food.

   _____

   _____

   c) The deer tick attaches itself to an animal and feeds on the animal's blood.

   _____

   _____

3. Invasive species are organisms that are not naturally found in a community. They are usually brought in by people, sometimes for a reason and sometimes by accident. In 1998, zebra mussels were accidentally introduced into Lake St. Clair. They quickly spread throughout the Great Lakes and into many inland lakes, rivers, and canals. They have nearly eliminated the native clam population in the ecosystem. What effect can an invasive species have on a habitat community? Why?

_____

_____

_____

_____

Adaptations

Organisms have characteristics that help them survive in their community. These adaptations can be physical structures or behaviors. Individuals with helpful traits are more likely to survive. They produce young with the same traits. Adaptations create biodiversity after many generations.

Structures

Physical features can help an organism survive. The beaks and feet of birds are good examples:

- short strong hooked beaks for eating animals
- long slender beaks dig in mud for food
- webbed feet for swimming
- a long back toe to perch on branches

a change in color is a type of adaptation

Structural adaptations can also be inside an animal. For example, a penguin's heart beats 60 to 100 times per minute. When the penguin dives, its heart rate drops to 20 beats per minute. That helps the penguin use less oxygen.

Behaviors

The things organisms do are called behaviors. They can help them find food or survive harsh weather. Hibernating and migrating are two behavioral adaptations. Hunting is another behavior.

Yummy fish swim into the shade made by a snowy egret's wings. Nice catch!

Brain Stretch

Write about why adaptations are important for an animal's survival. Use information from the reading and your own ideas.

"Adaptations"—Think About It!

1. Choose two wild organisms—one plant and one animal. These organisms must live in your community habitat or another habitat that you know well. For each organism, describe one adaptation that helps the organism survive.

_____
_____

2. Read the descriptions below. For each, tell how the adaptation helps the organism survive.

a) The American alligator digs a den in the mud in very hot weather.

_____
_____

b) A beaver can close its lips behind its front teeth.

_____
_____

c) The eyes of a hippopotamus are on top of its head.

_____
_____

3. Many organisms live in habitats that make survival difficult. These habitats include deserts, the Arctic, or the ocean floor. Design an imaginary organism that could survive in one of these habitats. Choose a habitat, then draw and label a picture of your organism below. Make sure your organism has adaptations that would help the organism survive. On another piece of paper, write a short paragraph to explain how the adaptations help the organism.

Chalkboard Publishing © 2012

How Biodiversity Benefits Humans

We get many things from other organisms. We get wood to build homes and get foods such as corn. Having many types of organisms gives us many resources.

Medicine

Humans have used plants as medicine throughout history. In ancient India, ginger was used for many things, including healing wounds and for stomach aches. The Gitxan peoples of British Columbia had many uses for a plant called Devil's club. The inner bark was used to heal wounds.

Plants are the starting point for many of the medicines today. The bark of the willow tree helps stop pain. An antibiotic comes from mold. The Pacific yew tree is the source of a cancer drug.

Clothes

Some cloth comes from plants. Flax turns into linen and cotton comes from cotton plants. Other fabrics come from animals. Silkworms make cocoons that we turn into silk. Wool comes from sheep, and leather comes from cows.

What Happens When There Is Less Diversity?

Less diversity can have a bad result. If there is only one type of corn, for example, bad weather could mean no corn grows. A single type of disease or pest could destroy the whole crop. But if there are many types of corn, some types could survive.

Growing only one type of plant takes out of the soil all the nutrients that plant needs. The soil then needs a lot of chemical fertilizer to grow more plants.

Some types of flowers will survive the pest that is eating the leaves.

Brain Stretch

Create a collage to show how humans benefit from biodiversity. Cut out pictures and words and paste them on a separate piece of paper.

"How Biodiversity Affects Humans"—Think About It!

1. On a piece of paper, list products your family uses that come from other organisms. Use a graphic organizer such as a web to classify the products. Create your organizer below. Share your organizer with a classmate.

2. A field grows one type of sunflower very well. Why might the farmer plant only that sunflower?

_____
_____
_____
_____

3. One type of grass does not grow well in dry ground. Another type does not grow well in wet ground. Why might growing both grasses together make a healthy lawn?

_____
_____
_____
_____

What Do You Think?

Think about the biodiversity in your city or community. What could be done to increase the biodiversity? You could plant a tree in your own backyard. Or an empty lot could be turned into a community garden. Talk about ideas with a classmate. Use your ideas to answer the questions below.

Think About It!

1. What could my family do? _____
_____
_____
_____
_____

2. What could our school do? _____
_____
_____
_____
_____
_____

3. What could groups of people do? _____
_____
_____
_____
_____

4. What could the local government do? _____
_____
_____
_____
_____
_____

Biodiversity Crossword Puzzle

Complete the crossword puzzle using words you have learned. On another piece of paper, use each word in a sentence that shows the word's meaning.

Across
2. class of animals with a spinal cord
7. animals without a backbone
10. an adaptation that is not a structure

Down
1. any living thing
3. characteristics of organisms that help them survive
4. things that move, reproduce, and grow are
5. the smallest part of an organism
6. different types of animals or plants together create this
8. the largest classification of organisms
9. a non-living thing that is all around you

All About Air

Air is all around you. You cannot live without air. Air makes flight possible too. Planes and rockets, birds and bumblebees—every type of flying thing uses air.

Some Properties of Air

Unless the air around you is heavily polluted, you cannot see, taste, or smell the air. If you think you cannot feel air, take a deep breath. You can feel air rushing into your lungs.

Air also has mass or weight. A balloon weighs more after you fill the balloon with air. You experience the pressure air exerts on every surface. Closer to the ground, more air is pushing down on top on you and Earth's gravity pulls more. The higher up you go, the less air pressure there is. Air becomes thinner and more spread out.

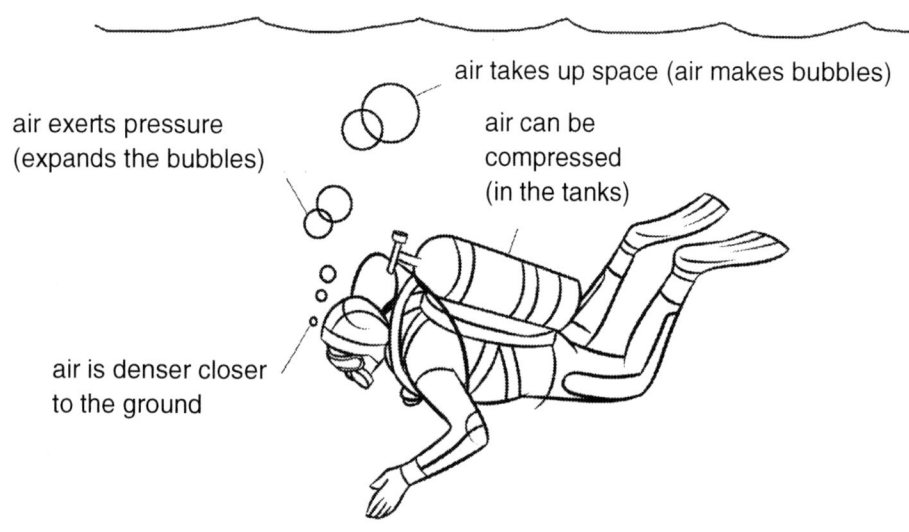

When it comes to flying, pressure is probably the most important property of air. Air pushes on every surface. Try this: Hold two pieces of paper out in front of you, about 3 in (8 cm) apart. Now blow between them. Did you expect the sheets to fly apart? They move closer together because the force of air on either side of the paper is unbalanced. The air moves quickly between the two papers. The moving air exerts less pressure than the air pressing on the outsides of the papers.

Aerodynamics means the motion of air and how air affects everything moving through it. People who design flying things have to know all about aerodynamics. This helps them build aircraft that fly efficiently and safely through the air.

"All About Air"—Think About It!

1. On another piece of paper, list all the things you know about air. For example, can air move? Can air move other things? Does air have a temperature? Can air rise or fall? Can air hold moisture? Write down how you know these things.

2. An experiment is shown in the diagrams below. The bottle contains only air. Diagram B shows what happened after the bottle was placed in hot water.

a) How does the hot water affect the temperature of the air in the bottle?

_____

_____

b) When the bottle is placed in hot water, the balloon expands. What two things happened to the air in the bottle? How do you know?

_____

_____

c) Based on this experiment, what property of air could be added to the article "All About Air"?

_____

_____

d) If the bottle in diagram B is placed back in the cool water, what will happen to the balloon? Explain why.

_____

_____

Taking Flight

Find out about the four forces of flight: lift, weight, drag, and thrust.

Just a Little Lift

Lift helps an airplane overcome its weight. Lift also helps the plane rise into the air, and stay up in the air. Many parts of a plane work together to create lift. Most lift is generated by the wings.

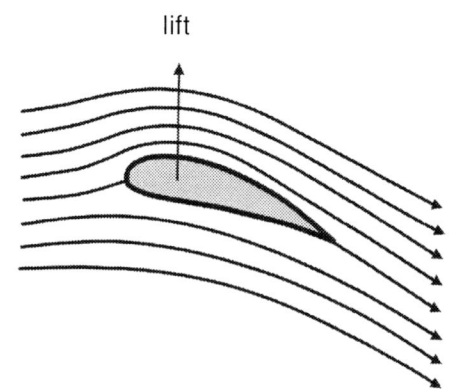

Look at a bird's wing when it is flying. You will see that it curves so that the top side is longer than the bottom. Planes create lift in the same way.

Air moving over the curved wing top goes farther than air going under the wing. The air also moves faster. Air that is moving more quickly exerts less pressure. So the air below the wing pushes up harder than the air above the wing pushes down. That lifts the plane into the air.

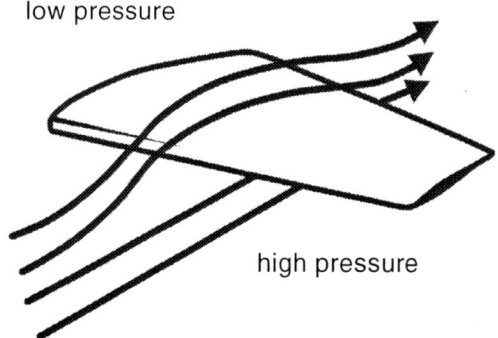

What a Drag!

When a plane moves through the air, the plane has to overcome drag. *Drag* is a force similar to friction. Drag slows down objects moving through air. Every part of a plane generates drag, even the engines.

The more streamlined, or aerodynamic, the plane is, the less drag the plane creates. The less drag the plane creates, the faster and more easily the plane can move through the air.

Overcoming Gravity

Earth's gravitational force makes it difficult for a plane to fly. You measure the force of gravity (weight) whenever you step on a scale. Gravity is a force that keeps a plane on the ground. The weight of plane has to be overcome before it can fly.

Give It Some Power

Thrust is the force that propels an aircraft through the air. Thrust is used to overcome the plane's weight and the drag of the air on the plane.

Birds create thrust by flapping their wings. In an aircraft, thrust is generated by the engines. Thrust makes lift possible.

"Taking Flight"—Think About It!

1. Think about how the four forces of flight are related. Use the words below to complete the sentences.

   drag   land   less   lift
   more   takeoff thrust   weight

   a) _____ must be greater than weight for a plane to _____ .

   b) Thrust must be _____ than _____ for a plane to take off.

   c) Lift must be less than a plane's _____ for the plane to _____ .

   d) _____ has to be _____ than the drag for a plane to land.

2. Identify the force in each example: drag, lift, weight, or thrust.

   a) A bird stretches out its wings for takeoff: _____

   b) Sinking into mud: _____

   c) Pushing with your foot to make a skateboard go faster: _____

   d) Using a parachute to fall through the air more slowly: _____

A Short History of Flight

As far back as the year 1500, the inventor and painter Leonardo da Vinci drew airplane-like machines. He never built any of them, but he inspired other inventors.

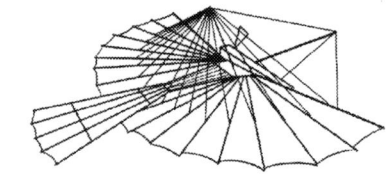

Hot-Air Balloons

Since about 250, hot-air balloons were used as military signals in China. In 1783, brothers Joseph and Étienne Montgolfier had the idea of sending people up in a hot-air balloon. No one knew if there was air that far above Earth's surface. And, if there was air, was that air safe to breathe? Onlookers were amazed and shocked when it worked.

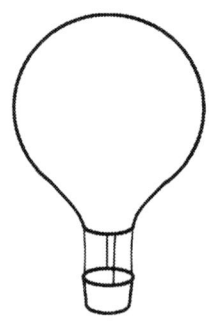

Bernoulli Principle

The hot-air balloon was a step forward in air travel, but the balloon could only drift in the wind. Inventors wanted a machine they could control. Serious inventors knew of the work of Daniel Bernoulli. In 1738, Bernoulli realized that the way air flows past a curved wing pushes or lifts it into the air.

Using Bernoulli's principle, British experimenter Sir George Cayley designed several gliders. A glider depends on wind to move and stay aloft. A glider has no engine, but it can be steered. In 1849, Cayley launched a glider that carried a 10-year-old boy a short distance. That was the first time a glider had flown with a person on board.

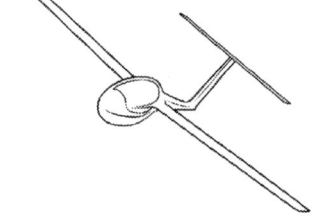

Cayley's work inspired Otto Lilienthal, a German engineer. By 1896, he had made about 2,000 glider flights. Lilienthal added an engine to power his glider, but it crashed, killing him.

Orville and Wilbur Wright

This tragedy frightened off many aircraft builders, but not American inventors Orville and Wilbur Wright. They were determined to add power to their gliders. In 1903, the Wright brothers became the first people to fly in a controlled and powered aircraft.

Better fighter aircraft were needed during World War II. In the early 1940s, airplanes became jet propelled. Today they are used to fly people for business or pleasure. They are also used to deliver goods around the world.

"A Short History of Flight"—Think About It!

1. Complete the timeline below to summarize information from "A Short History of Flight." (Use point form to list accomplishments.)

When	Who	Accomplishment

2. What do you think flight will be like 50 years from now? List your predictions on another piece of paper.

Flying Through the Sky

A car zooming along a road can move in just two different ways. A car can drive to the left or right and it can move up and down a hill. A plane can also tilt and twirl. Pilots use hinged flaps to change the shape of the plane in the air. This changes the aerodynamics and moves the plane.

The pilot can tilt the plane's nose up or down by tilting horizontal strips on the tail up or down.

Flaps on the wings tilt up and down to roll the plane.

A hinged rudder on the vertical tail makes the plane turn left or right.

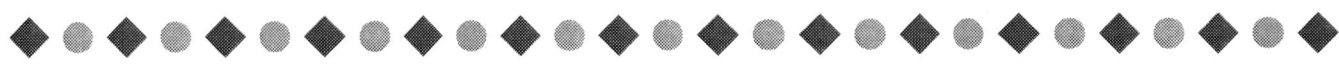

"Flying Through the Sky"—Think About It!

1. How does tilting a flap change the drag on a wing?

2. To go faster, should the flaps be flat or tilted? Explain why.

3. When a plane rolls, wind pushes up on one wing and down on the other wing. How does the pilot create these forces?

4. How could a bird change the tilt of its wings to slow down?

5. What force does the engine create?

Experiment: Build an Aircraft

What You Need

Materials such as
- Paper or plastic bags
- Pipe cleaners, straws, craft sticks
- Glue, tape, paper clips
- Measuring tape, timer

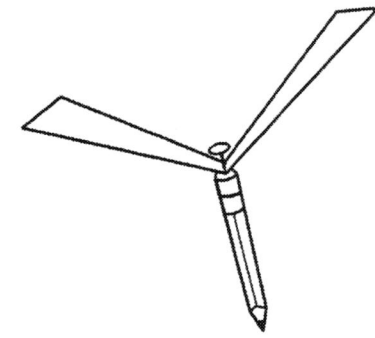

What You Do

1. Decide what type of aircraft you will build. For example, you might build an airplane, a spinner, or a glider.
2. Draw a plan and list the materials you need.
3. Decide how you will test the aircraft. Will you test the steering, time in the air, distance travelled, or all three?
4. Build the aircraft and conduct flight tests.
5. Record your findings.

Test	Flight Time	Distance Travelled	Steering
1			
2			
3			

Think About It!

1. Did your aircraft fly the way you thought it would? Explain.

_____

_____

2. What changes could make your aircraft fly better? Explain why.

_____

_____

Aircraft with Motors

Amphibious Aircraft

Aircraft that can land and take off either on land or water are amphibious. Some have skis so they can also land on snow or ice. Amphibious aircraft are especially useful in remote areas where they can land on runways, lakes, or rivers.

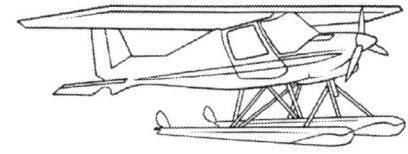

Helicopters

Helicopters get lift and thrust from the propellers—a rotor on the top. This rotor allows helicopters to take off and land vertically. It also lets the helicopter hover, and fly forward, backward, and sideways. Helicopters are often used in areas where there is no room for an airplane runway.

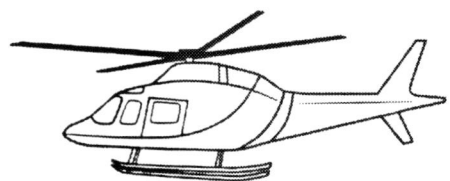

Stealth Planes

Stealth aircraft are military aircraft. These planes use advanced technology to avoid being detected by radar. Stealth planes have features that interfere with radar. These features make the plane difficult to see, hear, or detect with hi-tech equipment. However, these aircraft can be detected when they use their weapons. Stealth planes can have human pilots. They can also be flown by remote control.

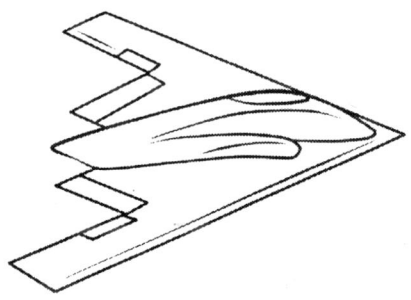

Think About It!

1. People are searching for a hiker lost in the wilderness. Give three reasons why a helicopter would be better to search with than an airplane.

_____
_____
_____
_____

Aircraft Without Motors

Some aircraft do not need a motor.

Hot-Air Balloons

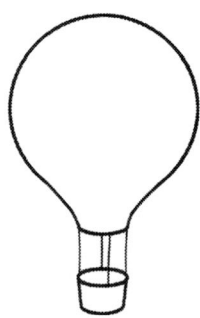

To take a ride in a hot-air balloon, you climb into the basket that hangs from the huge balloon. In the basket is a source of heat that warms the air in the balloon. This warm air makes the balloon rise. Hot-air balloons are used for recreation. They are also used to send weather recording devices high into the sky.

Kites

Kites are some of the oldest flying objects. Kites have been used as signals. They have also been used to take pictures from high in the sky, and to lift scientific instruments. Today, kites are mostly used for fun. Maybe you flew a kite at the park. Some people use kites many metres long to pull them across the water or snow.

Gliders

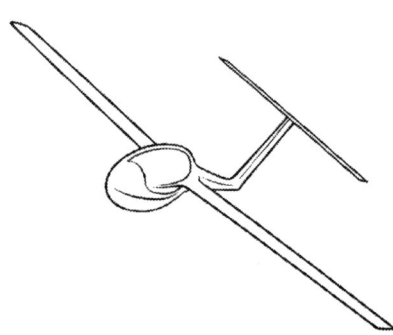

Because they have no engines, gliders depend on wind and air currents to keep them in the sky. Many gliders need a plane to get them up in the air. They are only designed to descend. Gliders are mostly used for sports such as gliding and hang gliding. They are also used for flight research. Paper airplanes that you make are actually gliders.

Think About It!

1. Hot-air balloons use very little fuel compared to other aircraft. What are two disadvantages of hot-air balloons?

_____

_____

2. In what ways is a kite similar to a glider?

_____

_____

Space Flight

In 1957, Russian engineers launched the first artificial satellite into space. A satellite is an object that orbits a planet or the Sun. This amazing event happened just 54 years after the Wright brothers' first flight.

One of the toughest obstacles to space travel is overcoming Earth's gravity. Doing this takes rocket power—extremely strong thrust from a rocket engine. These engines push rockets up by ejecting exhaust downward very quickly.

Sending people beyond the moon will be very difficult. How will they have enough oxygen, water, or food to survive the long trip out and back? People experience very little gravity in space. The human body has difficulty staying healthy in microgravity conditions.

Another obstacle to space travel is radiation. The Sun gives off many forms of radiation, including X-rays. This radiation penetrates spacecraft walls and astronauts' bodies. Earth's atmosphere helps protect us from this radiation. There is no protective atmosphere in space.

space shuttle

Brain Stretch

You have been hired to promote the importance of investigating space travel. On a separate piece of paper, draft a persuasive radio commercial. Use this checklist to help you. When you are finished, share your radio commercial with the class.

- ❑ My commercial is 15 to 30 seconds long.
- ❑ My commercial has a clear message about the benefits of space travel.
- ❑ I created the commercial to appeal to my target audience. (kids or adults)
- ❑ I practised reading my commercial with expression.

"Space Flight"—Think About It!

1. List the source of each force during a rocket launch.

   a) thrust: _____

   b) lift: _____

   c) drag: _____

   d) gravity: _____

2. What are some problems people will have to overcome for long-term space travel?

   _____
   _____
   _____
   _____

3. Satellites take photos from space and transmit signals for radio and phones. What disadvantage do satellites have?

   _____
   _____
   _____
   _____
   _____
   _____
   _____
   _____
   _____

communications satellite

Animal Fliers

You have learned a lot about machines that fly. Many animals can fly too. Some have wings. Some have feathers. Others stretch skin into a wing shape.

Insects

Insects are amazing fliers because they can turn their wings. That lets them hover in the air or even fly backwards. Spiders coast through the air on silky lines that catch the wind.

Bats

When you look at a bat's wing, you may see long bony fingers with a thin skin or membrane stretched between them. Bats flap their spread-out fingers to fly.

bat

Birds

Birds flap their wings to propel them through the air. They change the shape and angle of their wings to take off, steer, and land. Some birds are excellent gliders and barely flap their wings.

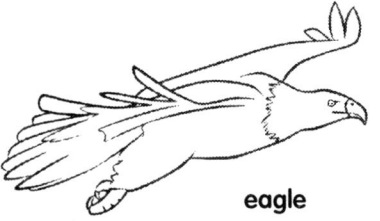
eagle

Think About It!

1. Animals that fly have very light skeletons. An animal of similar size that does not fly is much heavier. How does a light skeleton help a flying animal?

_____

_____

_____

_____

2. List a reason why each of the following are able to fly:

Insects: _____

Bats: _____

Birds: _____

Chalkboard Publishing © 2012

Furred and Finned Fliers

Squirrels

A flying squirrel can not really fly, but it is an excellent glider. It can coast farther than the length of three hockey rinks! First, a flying squirrel opens up a flap of furry skin that stretches from each wrist to each ankle. Then it glides with its body spread out. Its tail is puffed out so it works like a parachute.

flying squirrel

Reptiles

When a flying lizard unfurls its wings, it is easy to see the ribs that support the wings. The ribs and attached membrane spread out to form a semicircle on each side of the lizard's body. When these wings are not in use, they fold back against the animal's sides.

flying lizard

Fish

Flying fish cover long distances by making a series of glides through the air. At the end of each glide, a flying fish dips its tail into the water. This action produces new forward thrust. Flying squid propel themselves out of the water by expelling water. These creatures usually glide through the air only when they need to escape predators.

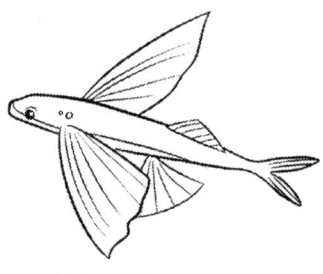

flying fish

Other Fliers

It is not only animals that use wings and air currents to get around. Dandelion seeds use their fluff like a parachute to float to new environments. Maple keys spin through the air like a helicopter or fall slowly. This helps them move away from the tree where they were produced.

Think About It!

1. Imagine a toy that would fly like an animal. What design features should it have? List two things.

_____
_____
_____
_____

Airplanes

"Airplanes"—Think About It!

1. Look at the picture on the previous page. In the chart, note the pros and cons of airplane travel. (Use point form.) Add any ideas you have.

Pros	Cons

2. How might these groups feel about airplane travel? Why?

   a) People who work in the aircraft industry or in airports: _____
   _____
   _____

   b) People who live very close to an airport: _____
   _____
   _____

   c) People who must travel to see doctors: _____
   _____
   _____

Flight Quiz

See how many of these questions you can answer before you look back at the articles in this unit or other resources.

1. What are the four forces of flight?

   _____
   _____
   _____
   _____

2. What property of air is illustrated by blowing up a balloon?

   _____

3. All the air in one container can be forced into a smaller container. What property of air does this show?

   _____

4. An inflated balloon weighs slightly more than it does when it is not inflated. What property of air does this show?

   _____

 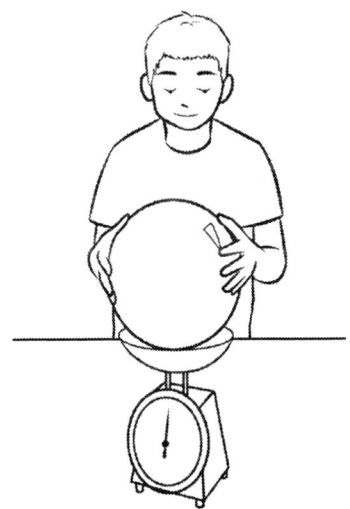

5. a) Air flows past a curved wing. Why does the air moving under the wing exert more pressure than the air moving over the wing?

_____
_____
_____
_____

b) Air pushes up on the wings. The pressure above the wing is not as strong as the pressure under the wing. What force is this?

_____

6. What effect does drag have on objects moving through the air?

_____
_____
_____
_____

7. What generates the thrust that helps a rocket take off?

_____

8. Drag is a force to overcome. How is drag useful to control flight?

_____
_____
_____
_____

Using Electricity

Electricity is a form of energy that we use in many ways every day. You probably use electricity much more than you realize. Look at the pictures. Circle the things that use electricity.

Electricity provides the energy that makes devices work. Some devices get electricity from batteries. Other devices get electricity from a wall outlet. You plug the device into the outlet and electricity travels into the device through the wire.

Where Does Electricity Come From?

Electricity is produced in a power plant by machines called generators. The generators are powered by a fuel such as coal or natural gas. Generators can also be powered by wind or water.

The electricity produced by the power plant travels over wires called transmission lines. Some of these transmission lines are attached to tall metal towers, and others are attached to poles along streets. In many places, electrical wires are buried underground. The wires provide electricity to homes, schools, factories, and other buildings.

"Using Electricity"—Think About It!

1. How do people use electricity at home? Rank the examples, starting with number 1 for the item that uses the most electricity over a year.

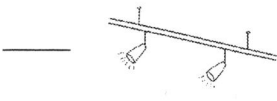 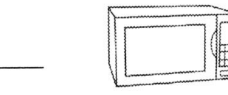 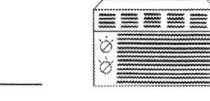

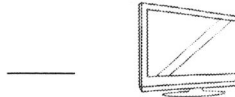  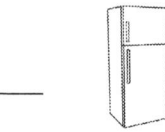

2. Electricity is used indoors and outdoors. List at least three ways that electricity is used outdoors.

_____
_____
_____

3. Many of the devices we use every day run on electricity from batteries. List at least four devices that use batteries. (Remember that any rechargeable device has a battery inside.)

_____
_____
_____

4. Electricity in wall outlets is not available during a power blackout. Describe how life would be different for you and your family during a blackout.

_____
_____
_____
_____
_____
_____

Current and Static Electricity

Static Electricity

Pull a wool hat off and your hair tries to stand on end. Walk across a carpet and you may get a shock when you touch something. Why does this happen?

When objects rub, a charge builds up on the surface. The charge can be positive or negative, like the poles on a battery. You cannot see the charge on an object, but you might see what the charge does. The charge makes hairs stand up and balloons stick to walls. The charge can also make a stream of water bend toward the charge.

Because the charge stays in place for a while, we call this charge static electricity. When the charge moves from one object to another, you may see—and feel—the spark.

Current Electricity

Current electricity that flows along a path of electricity called a circuit. A circuit connects the source of a electricity to a load that does something useful. The source might be a power plant, or it might be a battery. A lightbulb is one type of load.

Think About It!

1. Circle the examples that use current electricity.

    shock from walking on carpet   **flashlight**   **lightning**

    power lines in a house   **solar calculator**   **light switch**

2. Items with the same charge push apart from each other. Items with opposite charges are attracted to each other. Do these items have the same charge or opposite?

    a) A balloon sticks to a wall. _____

    b) Hairs push away from each other so hard that they stand straight up. _____

3. When clothes come out of the dryer, some may stick together. We call this "static cling" because static electricity causes the clothing to cling. Explain what happens to make the clothes stick together.

    _____

Experiment: Electric Cereal

Try this experiment to see static electricity at work.

What You Need

- Hard plastic comb
- Liquid soap
- Dish towel or paper towel
- 12 in (30 cm) of thread
- Piece of dry O-shaped cereal
- Wool mitten or sock
- Table
- Tape
- Balloon (optional)

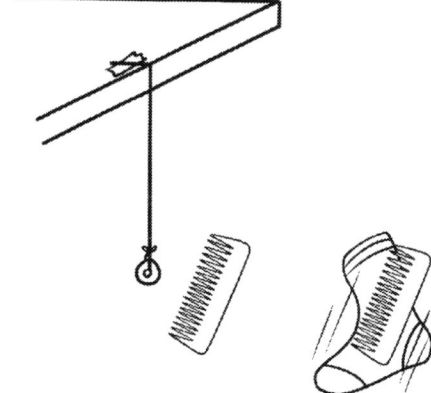

What You Do

1. Tie one end of the thread to a piece of cereal.
2. Tape the other end of the thread to the top of a table, so the cereal hangs in the air. The cereal should not be too close to anything.
3. Wash the comb with soap to make sure there is no oil on the comb. Carefully rinse away all of the soap. Dry the comb with the towel. (Make sure the comb is completely dry.)
4. Rub the comb quickly on the wool for several moments.
5. a) *Slowly* bring the comb close to the cereal. Do not touch the comb to the cereal. At a certain point, the cereal should move. Record what happens.
   b) Hold the comb still until the cereal moves again. Note how the cereal moves this time.
6. Once again, slowly bring the comb closer to the cereal. See if you can make the comb touch the cereal. Record what happens.

Optional

7. If a balloon is available, blow the balloon up and tie off the end. Repeat steps 4 through 6, using the balloon instead of the comb. Note any differences in the results.

"Experiment: Electric Cereal"—Think About It!

1. Record the results of your experiment in the chart below. Use drawings or words.

What I Did	What I Saw
Slowly moved the comb close to the cereal **(Step 5 a)**	
Held the comb still until the cereal moved again **(Step 5 b)**	
Tried to touch the comb to the cereal **(Step 6)**	

2. How does the comb become electrified?

_____

_____

3. Is this an example of static electricity or current electricity? Explain.

_____

_____

4. Did the comb and wool cause the effect you saw? Did the cereal cause the effect you saw? Describe a step you could add to test this.

_____

_____

Conductors and Insulators

Electricity flows through wires to make devices work. Electrical wires run through the walls in a house. Can electricity flow through all types of material? No. Some materials block the path of electricity.

A conductor is any material that electricity can flow through easily. Metals, water, and people all conduct electricity. Metal conducts electricity best.

An insulator is any material that electricity cannot flow through. An electric charge cannot move easily through rubber, plastic, glass, or wood.

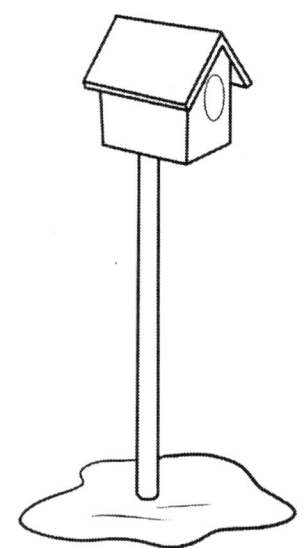

Conductors and Insulators in Daily Life

Electrical cords on devices such as televisions and lamps are covered in plastic. Inside the plastic is a metal wire. These metals wires are often made of copper. Copper is a good conductor.

Electricity flows through the wire, but not through the plastic. That means you can touch the cord without getting a shock.

When electricians work, they protect themselves by wearing thick rubber gloves. The gloves prevent electricity from flowing into the worker.

Brain Stretch

A finger can operate a touch screen. A tiny amount of electric charge in the touch screen moves into the finger. This tells the device where the finger is pointing. Do you think you could use a touch screen while wearing gloves? Explain.

"Conductors and Insulators"—Think About It!

1. After a big storm, you notice that one of the power lines on the street has snapped. One end of the power line is lying in a large puddle. You notice that the power line has a thick rubber coating on the outside. Would getting close to the snapped power line be dangerous? Would getting close be dangerous even if you did not touch the power line? Explain why or why not.

_____
_____
_____

2. Electricians often use tools that have rubber handles. Why is this a good idea?

_____
_____
_____

3. Imagine that you found an old lamp. The lamp's cord is covered in cloth, not plastic. Some of the cloth is falling apart. In one place, you can see the wire inside the cord. Explain why this is dangerous.

_____
_____

4. Categorize these materials using the Venn diagram below. Add your own examples.

pencil   coin   plastic button   stove element   rubber boots   safety pin
wood table   magnet   glass   touch screen   earphone wire   tires

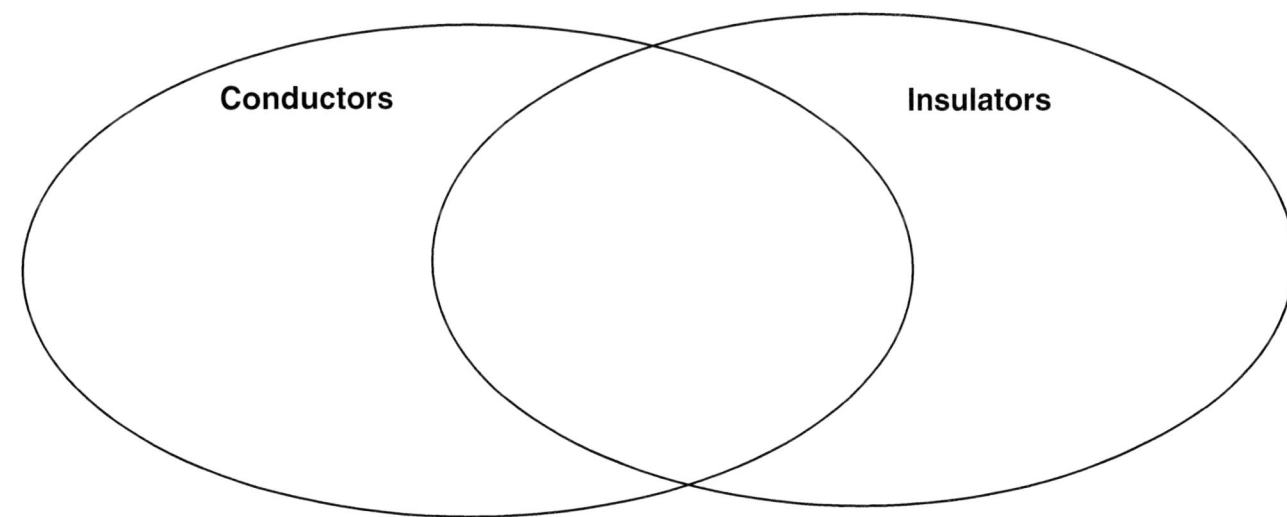

Using Water to Produce Electricity

Hydro means "water." Hydroelectric power plants use water to create electricity.

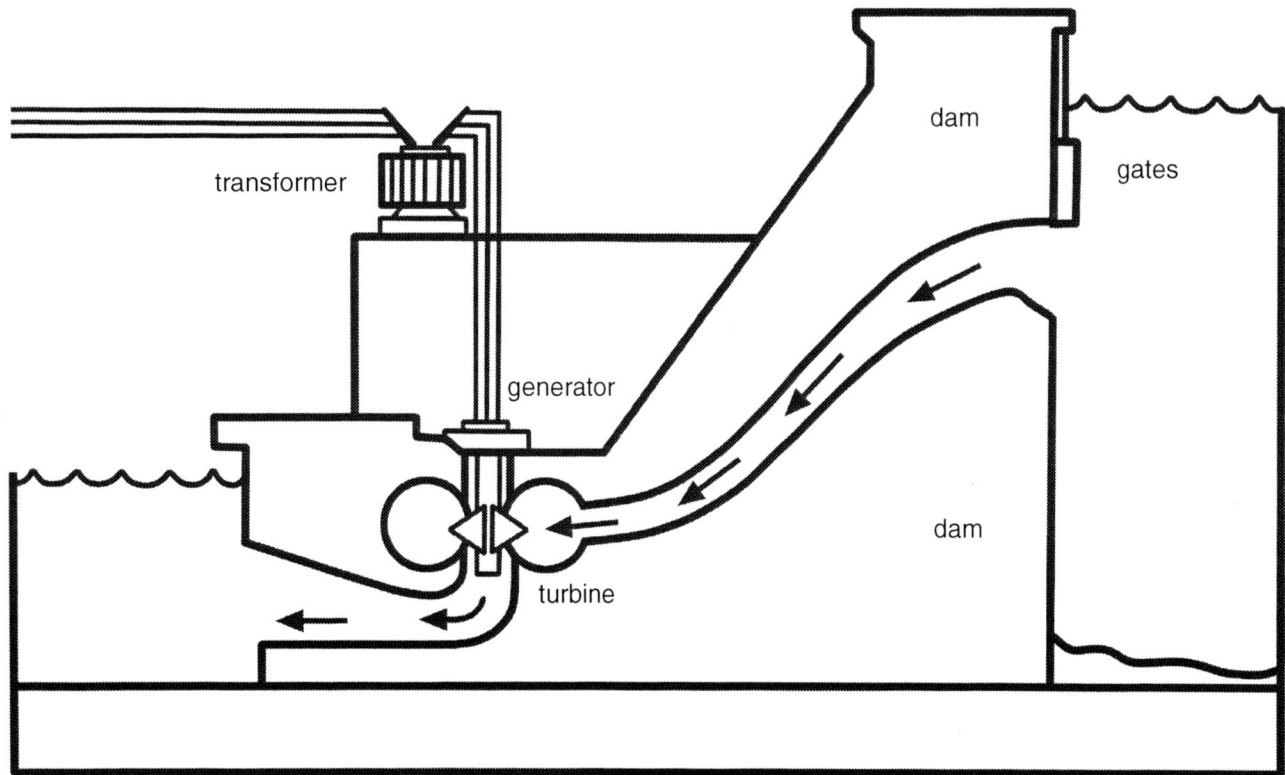

How It Works

1. A dam is built in a river. Water collects behind the dam, creating a lake or reservoir.

2. When the gates are open, water rushes down a pipe to the turbines.

3. The water turns the turbine blades, which turns the rod the blades are attached to.

4. The turbine makes the rod and magnets spin very quickly inside the generator. As the magnets spin, charge builds up in coils of copper wire, creating electricity.

5. The electricity travels along wires to the transformer. The transformer can help send electricity far away to where electrical power is needed.

"Using Water to Produce Electricity"—Think About It!

1. Explain how each group might feel about a hydroelectric dam being built.

   a) People who live along the edge of the river where the lake will be created

   _____
   _____
   _____

   b) People who enjoy boating and waterskiing

   _____
   _____
   _____
   _____

   c) Unemployed workers who live near the building site

   _____
   _____
   _____
   _____

   d) People who believe that an earthquake might happen where dam will be built

   _____
   _____
   _____

2. If you lived close to the site of the proposed dam, how would you feel about the dam? Explain why.

   _____
   _____
   _____

Using Wind to Produce Electricity

People have used wind power for centuries. Early windmills turned grinding stones that crushed grain into flour. Today we use windmills to create electricity.

How Wind Turbines Work

You can think of a wind turbine as the opposite of a fan. A fan uses electricity to make its blades turn, and the turning blades produce moving air. A wind turbine uses moving air (wind) to turn its blades. The movement of the blades produces electricity.

1. Wind turns the windmill's blades.

2. The blades turn the main shaft (rod) attached to them.

3. Gears in the gear box speed up the spinning motion. The second shaft then spins even faster. The faster the magnets spin, the more electricity is produced.

4. The turbine shaft and magnets spin very quickly inside the generator.

5. As the magnets spin, charge builds up in coils of copper wire. This creates electricity.

6. The electricity travels along wires to where it is needed.

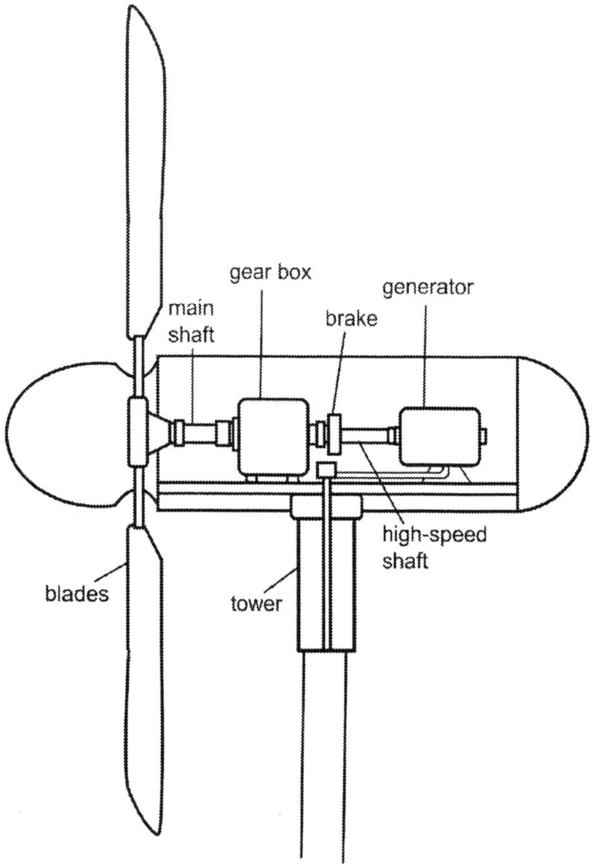

Wind Farms

A wind farm is a place that has a lot of wind turbines. The area must be flat and open, with a lot of wind. A wind farm can be built over water, such as a lake or ocean.

Advantages of Wind Turbines	Disadvantages of Wind Turbines
• create electricity • do not create pollution • many possible locations	• noisy • can be seen from quite far away • turning blades can kill or injure birds and bats

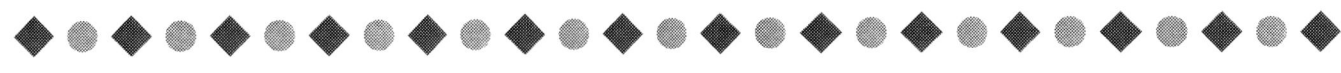

"Using Wind to Produce Electricity"—Think About It!

1. Why should wind turbine blades be very lightweight?

   _____
   _____
   _____
   _____

2. Wind turbines do not pollute the environment when they operate. But the metal, concrete, and other materials used to make them must be produced. The turbines must also be built and sent to the site. How can producing these materials affect the environment?

   _____
   _____
   _____
   _____

3. Review the information about hydroelectric power plants. In the chart below, list similarities and differences between a hydroelectric power plant and a wind turbine. (Use point form.)

Similarities	Differences

Transforming Energy

Energy exists in different forms. Electricity is one type of energy. Heat, light, and sound are also forms of energy. Motion is another form of energy.

You can change one form of energy into another. In fact, you do this every day. You turn motion into heat when you rub your hands together. A lightbulb turns electricity into light. A wall clock turns electrical energy into the motion of the hands.

Some devices give more forms of energy than we want. A blender makes motion (the blades turn), sound energy (noise), and heat (the motor warms up). Sound and heat energy do not help the blender accomplish the task of blending food.

A blender make motion by turning the blades.

Think About It!

Each device below turns electrical energy into another form of energy. Check off the forms of energy that each device uses to do its job. Add more devices to the chart.

Device	Transforms electrical energy into			
	Heat	Light	Sound	Motion
a) Television				
b) Iron				
c) Electric can opener				
d)				
e)				
f)				

All About Electrical Circuits

What Is a Circuit?

A circuit is the path that electricity travels along. Here are the parts of a simple circuit:

Part	What It Does	Example
Power source	provides electricity	battery
Conductor	carries the electrical current	metal wire
Load	uses electricity to do something useful	motor

The purpose of a circuit is to provide electricity to the load. If there is a gap anywhere in the circuit, the load will not work. The electric current will not be able to flow through the load. The load in the diagram is a lightbulb. If there is a gap in the circuit, the lightbulb will not light.

In the simple circuit, the electric current flows from the battery. The current flows through the lightbulb, and back into the battery. The ends of the two wires are connected to the battery and to the lightbulb. There are no gaps in the circuit, so the circuit is complete.

If there is a gap in a circuit, the electric current will not be able to flow through the circuit.

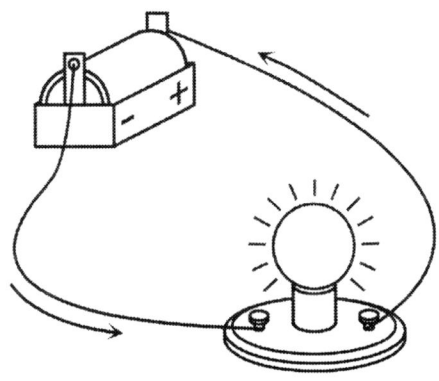
a simple circuit—closed

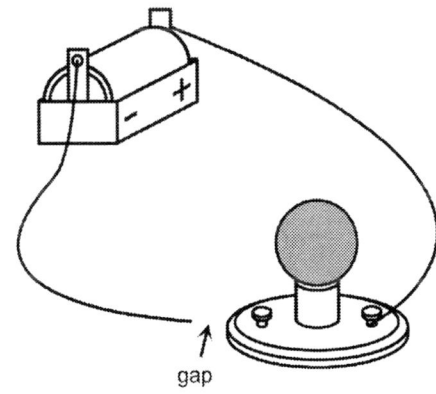
a simple circuit—open

What Is a Series Circuit?

A series circuit connects several loads. There is only one path for the electric current to flow along.

series circuit

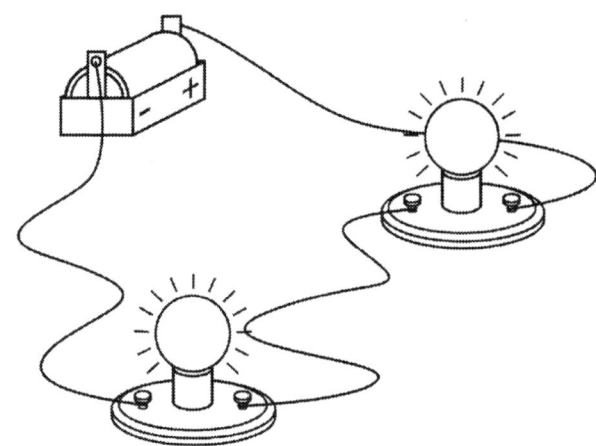

What Is a Parallel Circuit?

A parallel circuit provides more than one path for the electric current to flow on. Use your finger to trace these paths on the diagram:

Path 1: A → B → E → F
Path 2: A → B → C → D → E → F

parallel circuit

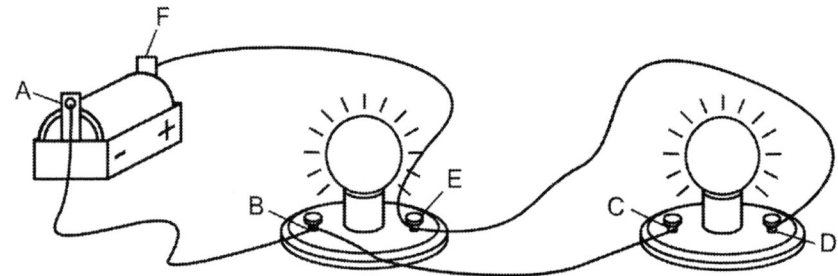

In a parallel circuit, the electric current flows through all the possible paths. This happens only if each path is complete. In the circuit above, some of the current follows Path 1 and some follows Path 2. Both lightbulbs light up.

"All About Electrical Circuits"—Think About It!

1. Incandescent lightbulbs light up when electric current passes through a thin wire inside the bulb. We say the bulb is "burnt out" when the wire has broken. In the circuit shown, lightbulb A is burnt out.

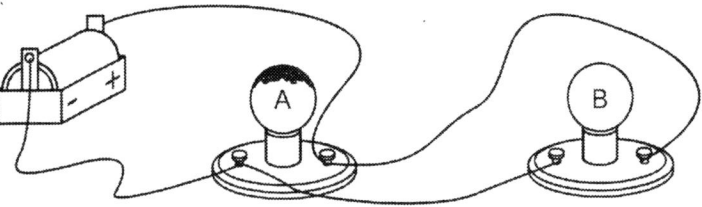

a) Is this a series circuit or a parallel circuit? Explain your answer.

_____
_____

b) Will lightbulb B light up? Explain why or why not.

_____
_____

2. When you use a switch to turn off a light, electric current no longer flows through the lightbulb. Use what you know about circuits to explain how a switch turns off a lamp.

_____
_____
_____

3. A string of patio lights has a parallel circuit. What are two benefits of this type of circuit?

_____
_____

4. a) Look at the diagram below. Is the mystery material a conductor or an insulator? How do you know?

_____

b) What could the mystery material be made of? Name three possible materials.

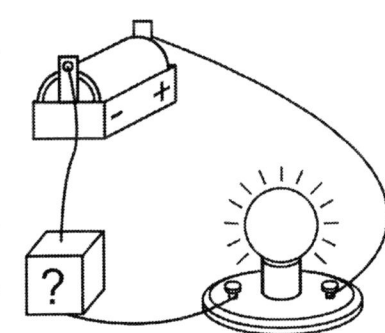

_____
_____

Life Without Electricity

If you could travel back in time, you would find that life was very different.

Cooking

The stove was heated by burning wood or coal. A lot of time was spent chopping wood to burn. The burnt wood or coal left ash behind. All that ash had to be cleaned out of the stove regularly.

Refrigeration

To keep food from spoiling, it was kept in an ice box. There were two doors on the ice box. The top door opened to small compartment where a big block of ice was stored. The bottom door opened to a larger compartment. Foods such as milk and butter were stored there.

In a city, a truck regularly delivered ice for the ice box. People in the countryside often cut blocks of ice from a frozen lake in winter. By covering the blocks with thick layers of hay or grass, they could make the ice last through most of the summer—if they were lucky.

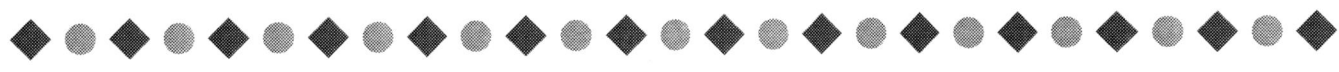

Lighting

Lamps that burned a fuel called kerosene were common. These gave off dirty smoke. You had to clean the lamps often, and add more kerosene.

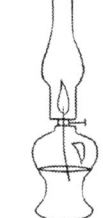

Warm Baths

The bath was moved into the kitchen to be close to the stove. Hot water was heated on the stove. Filling the tub took time, and heating water used up fuel. An entire family often took turns using the same bath water.

Laundry

Clothes were washed in soapy water and rubbed against a washboard to get them clean. The clothes were then rinsed in clean water. People did not do laundry on a rainy day because clothing had to hang outside to dry.

Ironing the wrinkles out of clothes was not much fun. The iron was made of solid iron. It was very heavy. The iron was heated on the stove. A rag or potholder was used on the very hot handle. The iron had to be reheated frequently on the stove.

Communication

There were no telephones or Internet. Messages were delivered by a person walking or riding a horse. Letters took days and sometimes weeks to reach their destination.

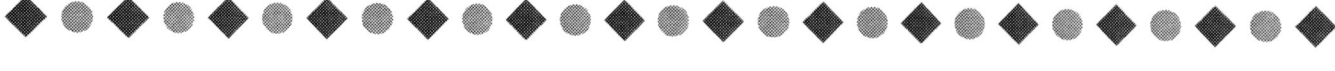

"Life Without Electricity"—Think About It!

1. Complete the chart below to show how we use electricity in daily life today. Then suggest how you might use less electricity for each task.

Task	How this task is done today	How I can save electricity
a) Cooking		
b) Refrigeration and freezing		
c) Lighting		
d) Warm baths		
e) Laundry		
f) Communication		

Electricity Crossword

Complete the crossword puzzle to review your knowledge of electricity.

Down
1. This type of electricity flows along a path
2. A material often used as an insulator on power cords
3. Type of material in which electricity flows easily
5. The part of a circuit that uses electricity to do something useful
6. Friction causes this to build up
8. You can see through this type of insulator
9. A small and portable power source

Across
2. Energy, such as electricity, is produced in this type of plant
4. Any material that stops the flow of electricity
7. The type of electricity caused by friction
8. The part of a hydroelectric power plant or wind turbine that produces electricity
10. This type of circuit has more than one path that electricity can flow along
11. In a series circuit with four lightbulbs, this is the number of bulbs that will light up when one bulb has burned out

Create an Electricity Game

What You Need

- Scissors
- Thumbtack
- Card paper
- Note paper
- Coloring materials
- Large paperclip

What You Do

Build a Spinner

1. Cut a large circle from the card paper. Divide the circle into 6 to 9 slices.
2. Decide on a category for each slice of the spinner. Make one category a question.

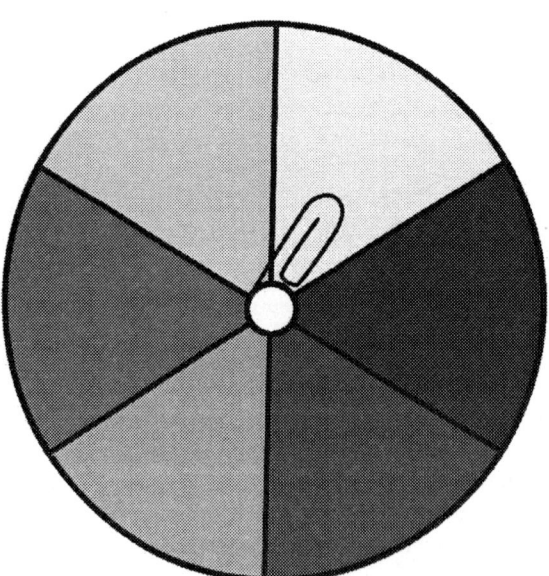

3. Use the thumbtack to secure the paper clip in the center of the spinner.

Write Question Cards

1. Divide sheets of notepaper into at least 20 question cards.
2. Look through your notes to write questions about the following:
   - electricity
   - circuits
   - power generation
   - safety
   - energy conservation

   Include the following types of questions:
   - true or false
   - multiple choice
   - explain something
   - draw a picture

3. Write questions on the cards. Include the answers and points.

Write the Rules

1. How many people can play? _____

2. Who keeps score? _____

3. What is the penalty for a wrong answer? _____

_____

4. When does the game end? It might be timed, last a certain number of turns, or have a winning score.

_____

Test your game with some classmates. After playing, adjust the rules and add questions to make the game better.

Our Sun Is a Star

Earth circles around the Sun. The Sun is a star and we could not live without it. The Sun is a ball of burning gas that gives us light and heat. The light makes plants grow, which provides us with food and with oxygen to breathe. The Sun's heat gives us rain by warming oceans and lakes. When that happens, some of the water evaporates into water vapor, which floats into the sky. When the water vapor rises high enough, it hits cooler air and condenses, and eventually falls as rain.

How big is the Sun? If the Sun were a hollow ball, one million Earths would fit inside it.

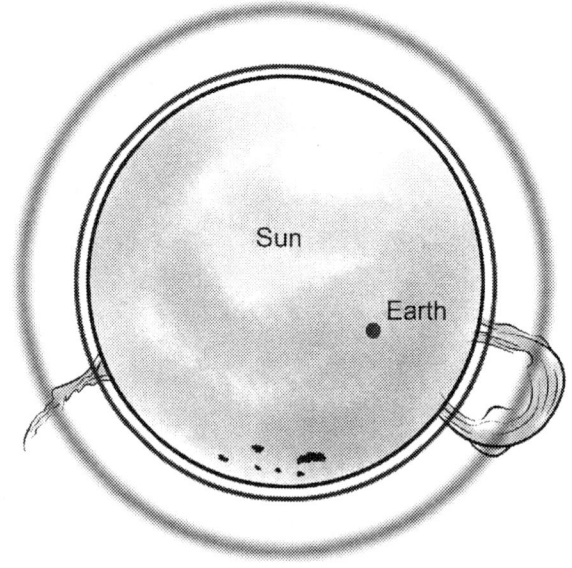

One million Earths could fit inside the Sun. The Sun has 333,000 times the mass of Earth.

Other Stars

Like the Sun, all stars are balls of burning gas. But stars are not all the same size and color. Stars can be 40 times as large as our Sun, or almost as small as Earth. Blue stars are very hot, while red stars are much cooler. In between are white and yellow stars. Our Sun is known as a yellow dwarf star.

The Sun seems larger than other stars in the sky because it is closer to us than any other star. Some stars are so far away that their light takes more than 1,000 years to reach Earth.

Think About It!

1. In what ways does the Sun affect Earth? Use the information from the article and your own ideas to write your answer.

   _____

   _____

Never look at the Sun! Even when you wear sunglasses, sunlight can damage your eyes.

Brain Stretch

All stars have a life cycle. Do some research to find out about the stages in the life cycle of a star. Create an information poster that includes diagrams and labels.

Experiment: Create a Sunset

The Sun looks yellow. Why are sunsets red? Do this activity to find out.

What You Need

- Clear glass bowl
- Water
- Spoon
- Few drops of low-fat milk
- White note paper
- Flashlight

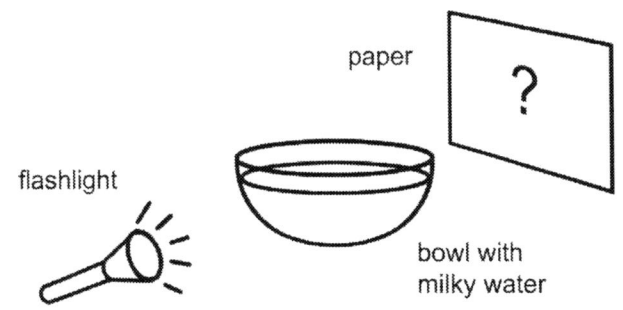

What You Do

1. Fill the bowl with water.
2. Hold up the piece of paper on one side of the bowl. Hold the flashlight on the other side of the bowl.
3. Shine the light through the water so that the light hits the paper. Record what you see.
4. Use the spoon to add a few drops of milk until the water looks cloudy.
5. Repeat steps 2 and 3. Record what you see.

Think About It!

1. What did you see when you shone the light through clear water? (Step 3)

   _____
   _____

2. What did you see when you shone the light through milky water? (Step 5)

   _____
   _____

3. How is the milky water similar to Earth's atmosphere?

   _____
   _____
   _____

Chalkboard Publishing © 2012

Planets in Our Solar System

There are eight planets orbiting our Sun. Each planet orbits the Sun on its own path. The planets' paths are hundreds of millions of miles (kilometers) apart. At any point in time, the planets are usually all at very different places along their paths. So, the distance between planets is even greater. Earth is so far from the Sun that the Sun's light takes eight minutes to reach our planet.

Each planet spins around as it orbits the Sun. On Earth, this creates night and day as different sides of the planet get sunlight. The time it takes for a planet to orbit the Sun is one year. Earth's year is 365 days long. Other planets have shorter years than Earth and some have much longer years.

Some planets have moons and some have rings. Some have atmosphere, winds, and storms. Huge craters, mountains, and canyons can be seen on some planets. Match each planet to a description below.

Mercury

Mercury is the closest planet to the Sun and is the smallest planet in our solar system. The surface is gray and rocky and has a lot of craters. With almost no atmosphere, it does not have storms, clouds, winds, or rain. With no atmosphere to trap the daytime heat, the temperature can drop 1,112°F (600°C) from day to night. No other planet has such an extreme temperature range.

Venus

Venus is about the size of Earth. The temperature on the surface is hot enough to melt lead. In fact, no planet is hotter than Venus. Not even Mercury, which is the planet closest to the Sun. Mercury and Venus are the only two planets that do not have moons orbiting them. Venus is easy to see. It reflects the Sun's light so brightly that some people mistake it for a star.

Earth

From space, Earth looks blue. That is because oceans cover 70 percent of Earth's surface. Earth is the third planet from the Sun and the largest of the four rocky planets. It is the densest planet. Earth spins on a tilt as it orbits the Sun. This tilt means that half the planet gets more solar energy than the other half. This tilt is the reason Earth has changing seasons.

Mars

Iron in the soil gives the planet a reddish color. Huge dust storms can almost cover the planet. Mars is about half the size of Earth, but has the highest mountain and the largest canyon of any planet. Space probes have looked for life on Mars, but found none. But scientists think the planet may once have been covered in rivers and oceans.

Jupiter

Jupiter is called a gas giant. This planet is so big that all other planets could fit inside it. Jupiter has at least 64 moons—more than any other planet. One moon is larger than Mercury. Jupiter is the fifth planet from the Sun, but can be seen easily from Earth. A space probe showed that Jupiter is circled by faint dark rings made of dust and bits of rock.

Saturn

No other planet has rings so bright. These rings are made of pieces of dust, rock, and ice. Some are as small as a fingernail, others are as big as a house. Some experts think the rings contains materials left over from when Saturn formed. Others think the rings contain pieces of nearby moons, chipped off by meteorites.

Uranus

Uranus is the only planet tilted on its side. Some experts think it may once have been hit by something huge that knocked it on its side. The planet is the third-largest in our solar system and is surrounded by faint rings. Unlike other planets, it has no cloud bands or storms. Uranus is so far from the Sun, that sunlight on Earth is 400 times brighter than the sunlight on Uranus.

Neptune

Neptune is a giant blue ball with wisps of white clouds. The winds blow up to 1,243 mph (2,000 km/h) on Neptune. No other planet has faster winds. Neptune is about the same size as Uranus and is the second-farthest planet from the Sun. There are four seasons on Neptune, but each season lasts more than 40 Earth years.

"Planets in Our Solar System"—Think About It!

1. Mars has two moons: Phobos and Deimos. How many moons do the four planets closest to the Sun have altogether? _____

2. Which planet has the highest-known mountain? _____

3. Which planet is the hottest? _____

4. Which planet is the smallest? _____

5. Which planet is the largest? _____

6. Which planet is the densest? _____

7. Which planet has the greatest daily temperature change? _____

8. Which planet spins on its side? _____

9. Imagine the Sun is the size of a basketball. Now match each planet with an item about its size. (Hint: three pairs of planets are about the same size as each other. For instance, Venus is almost the same size as Earth.)

   a) Mercury   i) small number cube
   b) Venus   ii) gum ball
   c) Earth   iii) peppercorn
   d) Mars   iv) large marble
   e) Jupiter   v) black bean
   f) Saturn   vi) kernel of corn
   g) Uranus   vii) pea
   h) Neptune   viii) small marble

Brain Stretch

Here is the order of the planets, starting with the planet closest to the Sun: Mercury, Venus, Earth, Mars, Jupiter, Saturn, Uranus, Neptune. The sentence below can help you remember the order of the planets. The first letter of each word is the first letter of a planet.

My Very Efficient Mother Just Served Us Noodles.

Moons

Did you know there are likely about 140 moons in our solar system? Jupiter has more than 64 moons. Mercury and Venus are the only planets that do not have a moon.

A moon can be any size. Two moons in our solar system are even bigger than Mercury. The second-largest moon is Titan, which orbits Saturn. Scientists are very interested in Titan because its atmosphere seems similar to Earth's long ago. Could there be life there now or sometime in the future?

Our Moon is about one-quarter the size of Earth. The surface is covered with boulders and a thick layer of gray dust. There are many large craters and mountain ranges, but no atmosphere. Millions of asteroids, comets, and meteorites have scarred the Moon's surface. There is no wind or rain to wash away these marks, so they remain unchanged for years.

The Moon Affects Life on Earth

The Moon circles Earth, reflecting light from the Sun and pulling on Earth's oceans. Ocean tides are caused by the Moon's gravity pulling the water toward the Moon. Since Earth rotates as this happens, two tides happen every day.

A lunar eclipse occurs when Earth comes between the Sun and the Moon. Earth's shadow makes the whole Moon look red and mysterious.

The Sun and Moon must be exactly opposite each other for a total eclipse. Partial eclipses happen much more often. A few lunar eclipses happen each year. Most can only be seen from certain places on Earth, depending on where the darkest shadow passes.

Phases of the Moon

Do you ever wonder why the Moon looks so bright in the night sky? Earth's Moon is not a star, yet it gives off its own light. This is because the surface of the Moon reflects the light from the Sun. The part of the Moon facing the Sun is lit up. The part of the Moon facing away from the Sun is dark. The sunlit part of the moon that we see is what is responsible for the Moon's phases. The phases of the Moon depend on its position in relation to the Sun and Earth. As the Moon makes its way around the Earth, the phases of the Moon change.

"Moons"—Think About It!

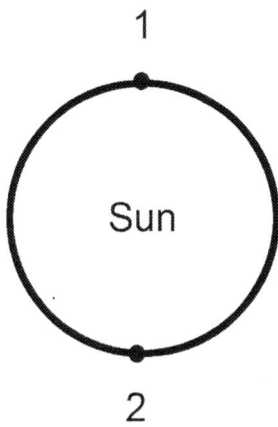
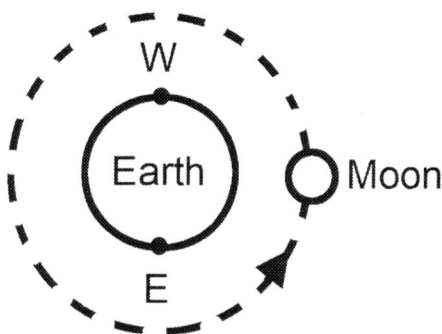

1. Draw a lunar eclipse. On the diagram above, use one color to join point 1 to points E and W. Extend the lines to the Moon. Color between the lines. Use another color to join point 2 to points E and W. Extend the lines to the Moon and color in the shape. Look at the shadow you colored between Earth and the Moon. The area of mixed color is the location of the darkest shadow.

2. Look at the Moon every few nights for about a month. Look in the newspaper to find out what time the Moon rises each night. Note what you see.

   a) The Moon does not actually change shape. Why does what you see change?
   _____
   _____

   b) Number the moon pictures below in the order that you saw them.
   _____

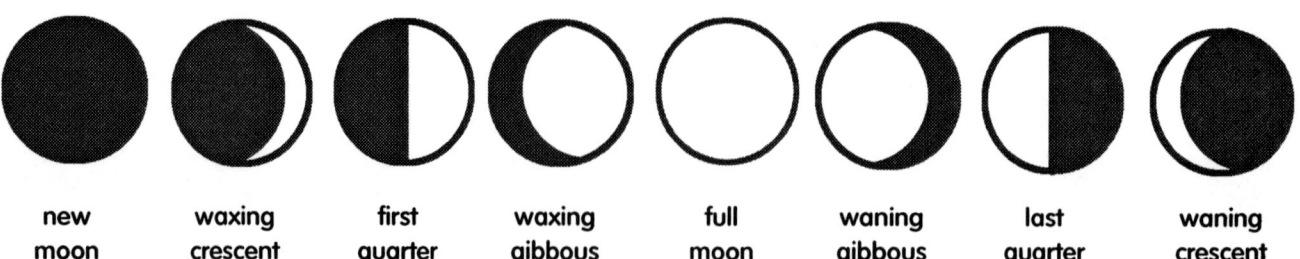

new moon | waxing crescent | first quarter | waxing gibbous | full moon | waning gibbous | last quarter | waning crescent

They Came from Outer Space

Comets

You might say a comet is a dirty snowball that orbits the Sun. This snowball is about the size of a small city! As this snowball zooms closer to the Sun, its surface begins to turn into gas. Gas and dust stream out into a tail that can spread out as far as 49,710 mi (80,000 km).

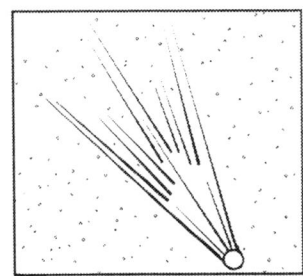

Comets mostly stay farther away from the Sun than our planets are. You will rarely see a comet unless you use a telescope or binoculars. Halley's Comet is the most famous comet. It swings by Earth about every 75 years. You will have to wait until 2062 to see it again.

Asteroids

These space rocks may be shaped like balls, lumpy slabs, or bricks. You find asteroids circling the Sun in the Asteroid Belt, a band that lies between the planets Mars and Jupiter.

The biggest asteroid is bigger than many American states. One asteroid even has its own moon. Most asteroids are only about the size of a house. You need binoculars to seem them.

Meteoroids

Small chunks of asteroids or old comets may become meteoroids. When these rocky bits enter Earth's atmosphere, they become meteors. Ever seen shooting stars? Then you have seen meteors. They burn up and shoot out fiery trails as they fall. If a meteor hits our planet, it gets a new name: meteorite.

Meteors are tiny. Most are as small as grains of sand and few get as big as baseballs. At certain times of year, many meteors streak through the night sky in a meteor shower. You can see them just by gazing up on a clear, dark night.

Think About It!

1. Arrange these in order of size: asteroid, meteoroid, comet.

2. Name one meteor shower and explain its name.

Experiment: Mass of Asteroids

Find out whether metal asteroids are heavier than stone asteroids.

What You Need

- A stone a little smaller than a golf ball
- 200 pennies
- 1 c (250 mL) measuring cup
- Warm water
- 2 small yogurt containers

What You Do

1. Put the stone in the cup and pour in water up to the 1 c (250 mL) mark.
2. Remove the stone without spilling any water. Place the stone in one of the yogurt containers.
3. Record the level of the water.
4. Add pennies to the water until the level is back up to the 1 c (250 mL) mark. Now you have measured a volume of pennies equal to the rock.
5. Carefully pour out the water and place the pennies in the other yogurt container. Now the containers hold the same volumes of stone and of metal.
6. Hold a container in each hand. Record which container is heavier.

Think About It!

1. Calculate the volume of the stone: 1 c – new level of water = volume of the stone.

   _____

2. Which was heavier: the stone or the metal? _____

3. What does this tell you about stone and metal asteroids?

   _____
   _____

4. How do you think the mass of a comet compares? Explain your thinking.

   _____
   _____

Explore the Night Sky

On the clearest and darkest night, you might be able to see 2,500 stars in the sky. You might also see a cloudy path across the sky. That is the faint glow from the other 300 billion stars in our Milky Way galaxy. You can also see some of the planets in our solar system. With binoculars or a telescope, you may even catch a glimpse of the Andromeda galaxy.

Venus is easy to see in our sky. It reflects the Sun's light so brightly that people can mistake it for a star. Mars and Jupiter can be easy to spot—sometimes only the Moon and Venus are brighter in the night sky. Look for Mercury early in the evening or in the early morning, when the Sun is below the horizon.

Have you ever imagined that the clouds look like animals? Sometimes people connect the dots and imagine that the stars make shapes too. Many cultures have traditional stories about the shapes they imagine in the starry sky. Some stories are about magical animals. Others are about ancient battles, or gods.

Think About It!

1. a) Go outside on a starry night and draw what you see.

Date:_____ Time:_____ Direction:_____

b) Use a star chart to identify what you saw. Add labels to your drawing. What planets did you see? Did you see any satellites, such as the International Space Station?

c) Compare drawings with a classmate. Did you both see the same stars and shapes? Explain any differences.

2. a) Find someone to tell you a story about the stars. Or, ask a librarian to help you find stories from around the world. Record the story. You might write it down or use a camera or voice recorder.

b) How many different stories about the stars did your class collect? Create a podcast or puppet show to share your favorite stories.

Astronauts and Space Travel

In space you do not feel gravity. It is extremely cold, and there is no atmosphere to breathe or to protect you from the Sun's radiation. It takes enormous energy to get there, and you have to take everything you need with you—even the air. Outer space is an extreme environment.

The first person in space orbited Earth for less than two hours in 1957. Less than 40 years later, another astronaut spent a record-setting 438 days in space. Space travel has changed a lot since then. Between 1969 and 1972, 12 astronauts set foot on the Moon. These are the only times a human has landed anywhere other than Earth.

After leaving Earth, life support is the most important function of a space craft. It must clean the air for astronauts to breathe and provide heat. It must also protect the contents from the X-rays and other radiation coming from the Sun. All things used in space are designed for the low gravity. Tools are attached to surfaces so they do not float away. Even drinking becomes difficult. You do not think about it, but gravity keeps the drink in your glass. Without gravity, the drink floats in a ball. Meals must be eaten from containers that keep the food from floating away. Equipment must be protected from floating crumbs too.

Gravity on Earth also makes your muscles work hard. This exercise helps the blood spread throughout your body. In space, blood does not circulate as well. Lack of exercise makes the body lose muscle and bone mass. Astronauts have to strap themselves into exercise equipment to keep fit without floating away!

Brain Stretch

What do you think are four character traits someone needs to have to become an astronaut? Justify your thinking for each character trait.

_____
_____
_____
_____

"Astronauts and Space Travel"—Think About It!

1. List 10 items you would need in space. Give a reason why you want to take each item.

_____
_____
_____
_____
_____
_____
_____
_____
_____
_____

2. What are the most important features to have on a spacecraft? Remember, the spacecraft has to provide everything an astronaut will need while in space.

_____
_____
_____
_____
_____
_____

Brain Stretch

Some of the things we use on Earth were invented for space travel. Heat cameras, insulation, multivitamins, and cordless tools are just a few examples. Use books or Internet resources to learn about one such invention. Create a demonstration to show people how to use it on Earth.

Other Spacecraft

Satellites

Space travel began with a satellite called Sputnik (Russian for "travelling companion"). This satellite was launched in 1957. It provided information about Earth's atmosphere. Today there are about 3,000 satellites orbiting Earth. They carry equipment that connects phones, television, and the Internet. Some satellites watch the weather, some send data to GPS units, others take pictures of Earth and the stars. One big satellite is a telescope.

Probes

The space probe Pioneer 10 was launched in 1972. It was the first spacecraft to fly close to Jupiter. The probe sent back the first close-up images of Jupiter, then continued out of our solar system. Pioneer 10's last signal was received in 2003. The probe is on a 2-million-year journey to a star.

Some probes land on planets. Probes such as rovers move, but stationary landers do not move. In 1997, the rover Sojourner landed on Mars and became the first rover to land on another planet. For more than two months, it sent back images of Mars. It also sent information about chemicals in the rocks and soil, and about the weather on the red planet.

Space Telescopes

Space telescopes are another way scientists can learn about the solar system. The Hubble Telescope was launched into orbit around Earth in 1990. The telescope has beamed back information about the universe's past and future. Hubble takes clearer pictures than a telescope on Earth because there is no atmosphere in space to warp the image. This telescope is solar powered and can be repaired in space.

International Space Station (ISS)

This is the largest and most expensive spacecraft ever built. The space station is one of Earth's artificial satellites, but it is also home to crews of astronauts. It is so big that you can see its glow as it crosses the night sky. The first part of the ISS was launched in 1998. Now it has solar panels for power, living quarters, and science labs.

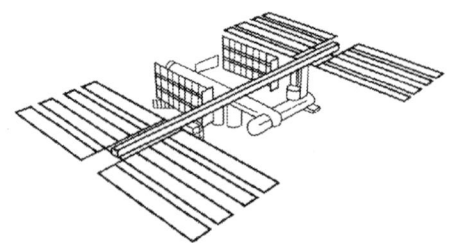

International Space Station (ISS)

Space shuttles carry food, water, equipment, and crew members back and forth between Earth and the ISS. The United States, Russia, Canada, Japan, and some European countries have all contributed to the station. The space station may one day be a launching base for a mission to the Moon, Mars, and beyond.

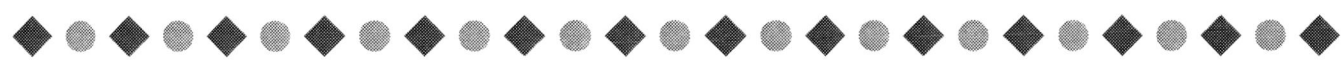

"Other Spacecraft"—Think About It!

1. What are the pros and cons of orbiting spacecraft and spacecraft that land on a planet?

   _____
   _____
   _____
   _____

2. Imagine designing a land rover to explore a new planet. List three things you have to plan for in the design. Describe two things you will need the rover to do.

   _____
   _____
   _____
   _____

3. A lander is a spacecraft that lands on another planet. What advantages does a rover have over a stationary lander?

   _____
   _____
   _____
   _____

> **Brain Stretch**
>
> Do you then think you see the lights of an airplane in the sky at night? It may be the International Space Station (ISS). You can often see satellites in the night sky. They reflect sunlight just like the planets do.
>
> Go outside just after sunset or right before sunrise. Those two times are when the ISS is lit by the Sun but you are in darkness. You cannot see the ISS at other times. Look online to get more information about the best place to look for the ISS.

Astronaut Research Report

Research one of these famous astronauts. Write a report about their early life and achievements. Include some interesting facts.

- Buzz Aldrin
- Kathryn Sullivan
- Ellen Ochoa
- Yuri Gagarin
- Neil Armstrong

- Guion Bluford, Jr.
- Mae Jemison
- Alan Shepard
- Edward H. White
- Eileen Collins

- John Glenn, Jr.
- James Lovell
- Sally Ride
- Franklin Chang-Dìaz
- Ellison Shoji Onizuka

Neil Armstrong

Name _____

Date of Birth: _____

_____
_____
_____
_____
_____
_____
_____
_____
_____
_____
_____
_____
_____

Space Technology

Space travel has changed your life in ways you probably never imagined. From your sunglasses to the shoes you wear hiking, space technology touches your life.

How were these devices first used in space?

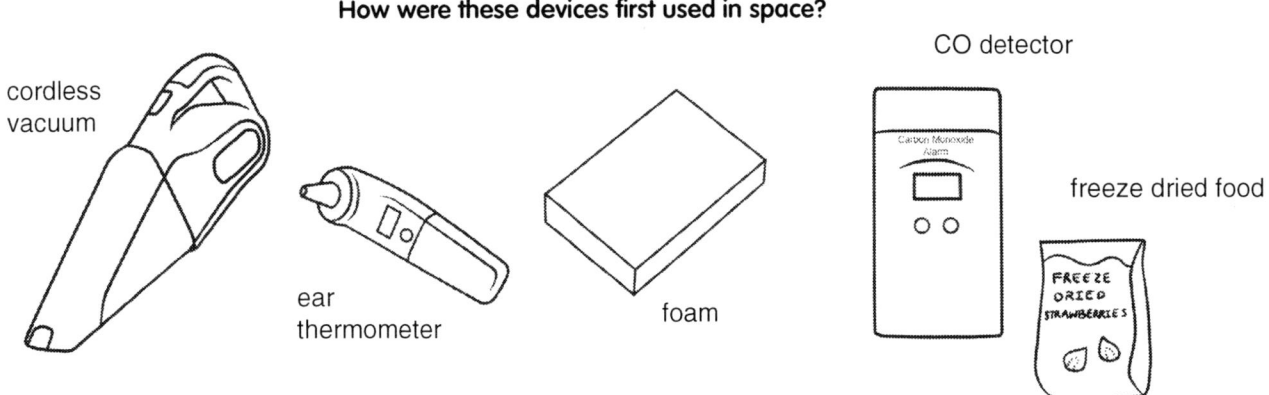

You likely already know that radio and television signals are transmitted around our planet by satellite. Other satellites help with navigation. Many cars now have global positioning systems (GPS), which are navigation systems that use satellites. Other products include the scratchproof coating on sunglasses, and medical technology. Insulin crystals (to fight diabetes) were grown in space. The technology used for space shuttle fuel pumps is now used in artificial hearts.

Space technology is great but it costs a lot. It takes an incredible amount of money to put an astronaut into space. Just a toilet on the ISS costs $19 million. Also, astronauts' dirty clothes cannot be cleaned, so these expensive garments are just thrown away. Is this the best way for a government to spend its money?

Here on Earth, money is desperately needed to improve medical care and other services. We also need more money to help the environment. Is it right to spend money on space research when the money is needed elsewhere? Are space programs just a way for one country to pretend it is better than another because it has more advanced space technology?

Space travel also creates a lot of pollution. Launching a spacecraft uses a lot of fuel. On Earth, people are trying to save fuel. Space research also adds to the garbage orbiting Earth or eventually landing on our planet. Some pieces are as large as a car. Even bits as small as flecks of paint can damage other spacecraft because of the speeds at which they travel.

The worst outcome of space research can be disastrous. Although every aspect is carefully planned, sometimes tragedies happen. In 1986, the Space Shuttle *Challenger* blew up shortly after takeoff. Then in 2003, the shuttle *Columbia* broke apart as it re-entered Earth's atmosphere. In both cases, all seven crew members were killed.

"Space Technology"—Think About It!

1. a) How do you feel about space technology and exploration?

_____
_____
_____
_____

b) What are the benefits? Do the benefits outweigh the costs? Explain your thinking.

_____
_____
_____
_____
_____

2. Do you think life on Earth would be better or worse without space travel? Explain your thinking.

_____
_____
_____
_____
_____

3. Imagine exploring the ocean in a submarine. How is this similar to life in space?

_____
_____
_____
_____

continued next page

"Space Technology"—Think About It! (continued)

4. Space technology has changed the foods we eat. There are six main ways food is stored for astronauts to eat on board the space shuttle. Match each method with its description.

a) **Fresh**

b) **Intermediate Moisture**

c) **Irradiated**

d) **Natural**

e) **Rehydratable**

f) **Thermostabilized**

i) Food is freeze-dried to remove water. Water is replaced before eating.

ii) Food with some water removed; water is not replaced before eating.

iii) Food is cooked and packed in foil pouches. The food is sterilized by radiation so it can be kept at room temperature.

iv) Food is heated to kill bacteria so it can be stored at room temperature.

v) Food is ready to eat and stored in flexible pouches.

vi) Food is not preserved so it must be eaten quickly.

5. Tell how you use space technology in your daily life.

_____

_____

_____

_____

Did You Know?

The robotic Canadarm was a remote-controlled manipulator that was attached to a space shuttle. There, it could capture, repair, and launch satellites, position astronauts, help hold equipment, and move cargo. The jointed Canadarm was launched in 1981. It could move objects weighing as much as 66,139 lb (30,000 kg). In 2001, Canadarm2 was launched. The new device played an important role in the building of the International Space Station (ISS).

Animals in Space—Fair or Not?

Cats, dogs, frogs, guinea pigs, and monkeys. Did you know all of these animals have been launched into space?

Before sending anyone into space, no one knew how travelling so far from Earth's surface would affect humans. Could astronauts survive the launch and re-entry? What about surviving space radiation? What effect would microgravity have on them?

Scientists knew that there were no machines they could launch that would give them all the answers they needed. They decided to send other animals to test the trip.

In 1957, Russian scientists launched the dog Laika (the name means "Barker" in Russian), who became the first animal to orbit Earth. A dog was chosen because scientists felt it could endure the long stretches of inactivity involved in space travel. Sadly, Laika died when her spacecraft overheated.

Today, animals are still sent into space to take part in experiments. For instance, does floating confuse animals? Laura Lewis, from NASA Ames Institutional Animal Care and Use Committee. tested that question. She found that mice adapted very quickly to floating. Within minutes, the mice were grooming themselves and eating.

Think About It!

1. Do you think it is acceptable to send animals into space? Why? If not, are there any times when you think it might be alright? If so, explain them.

_____
_____
_____
_____
_____
_____
_____
_____

Those Puzzling Planets

Solve the clues to complete the crossword on the next page.

Across

2. A natural satellite
4. Path made by something in space as it circles another object
5. The red planet
7. Planet known for its huge rings
10. Planet mostly covered in water
11. Seasons on this planet last 40 Earth years
13. Planet that spins on its side
14. Lack of atmosphere gives this planet extreme daily temperatures
17. Biggest object in our solar system
18. Objects orbiting the Sun that can be as large as a province

Down

1. Made up of the Sun and planets
3. Made by connecting the dots of stars in the sky
6. Its ice melts into a huge tail that streaks across the night sky
8. These showers are named for the constellation you see them near
9. Planet with strong acid rain
12. An enormous gas planet
15. Famous Canadian part of the space shuttle
16. The Milky Way is one

continued next page

Brain Stretch

The order of the planets outward from the Sun is Mercury, Venus, Earth, Mars, Jupiter, Saturn, Uranus, Neptune. To remember this, make up a sentence in which each word starts with the first letter of each planet, in order.

Those Puzzling Planets continued

Use the clues on page 81 to complete this crossword.

Science Brochure

A brochure is a booklet or pamphlet that contains descriptive information. Choose a science topic you have studied, or that you are interested in.

STEP 1: Plan Your Brochure	Completion Date
1. Take a piece of paper and fold the paper the same way your brochure will be folded.	
2. Before creating the brochure, plan the layout in pencil. • Write the heading for each section in the place where you would like it to be in the brochure. • Leave room underneath each section to write information. • Also leave room for graphics or drawings.	

STEP 2: Complete a Draft	Completion Date
1. Research information for each section of your brochure.	
2. Read your draft for meaning, then add, delete, or change words to improve your writing.	

STEP 3: Final Editing Checklist

- ☐ I checked the spelling.
- ☐ I checked the punctuation.
- ☐ I checked that sentences were clear.
- ☐ My brochure is neat and organized.
- ☐ My brochure has pictures or graphics.
- ☐ My brochure is attractive.

Oral Science Presentation Checklist

Use this checklist as you create an oral presentation on a science topic.

Science Topic: _____

How long does the presentation need to be? _____

Introduction Checklist

❑ I introduced my topic in an attention-grabbing way, such as
  ❑ a quote
  ❑ a statistic
  ❑ an example
  ❑ a question
❑ I state what I am going to talk about in 1 to 3 sentences.

Body Checklist

❑ Each key point has supporting details, examples, or descriptions.
❑ I wrote out my ideas the way I would sound if I were if I were explaining, showing, or telling someone in person during a conversation.
❑ I read aloud what I wrote.
Tip: You do not have to use full sentences. Write it the same way you talk.

Conclusion Checklist

❑ I summarized my key points.
❑ I ended my oral presentation in an attention-grabbing way, such as
  ❑ a quote
  ❑ a statistic
  ❑ a question

Presentation Delivery Tips

- Practise! Practise! Practise! Get comfortable with what you have written.
- Highlight your good copy in places where you would like to pause for effect, or emphasize a point.
- Think about hand gestures and making eye contact with the audience or camera.
- Think about your tone of voice to show enthusiasm, emotion, or volume.

Take One Point of View

Write an article that gives your point of view about a science-related issue. Use this outline to plan your article. Complete your work on a separate piece of paper.

A Statement of Your Point of View	
Main Idea	**Supporting Evidence**
Main Idea	**Supporting Evidence**
Main Idea	**Supporting Evidence**

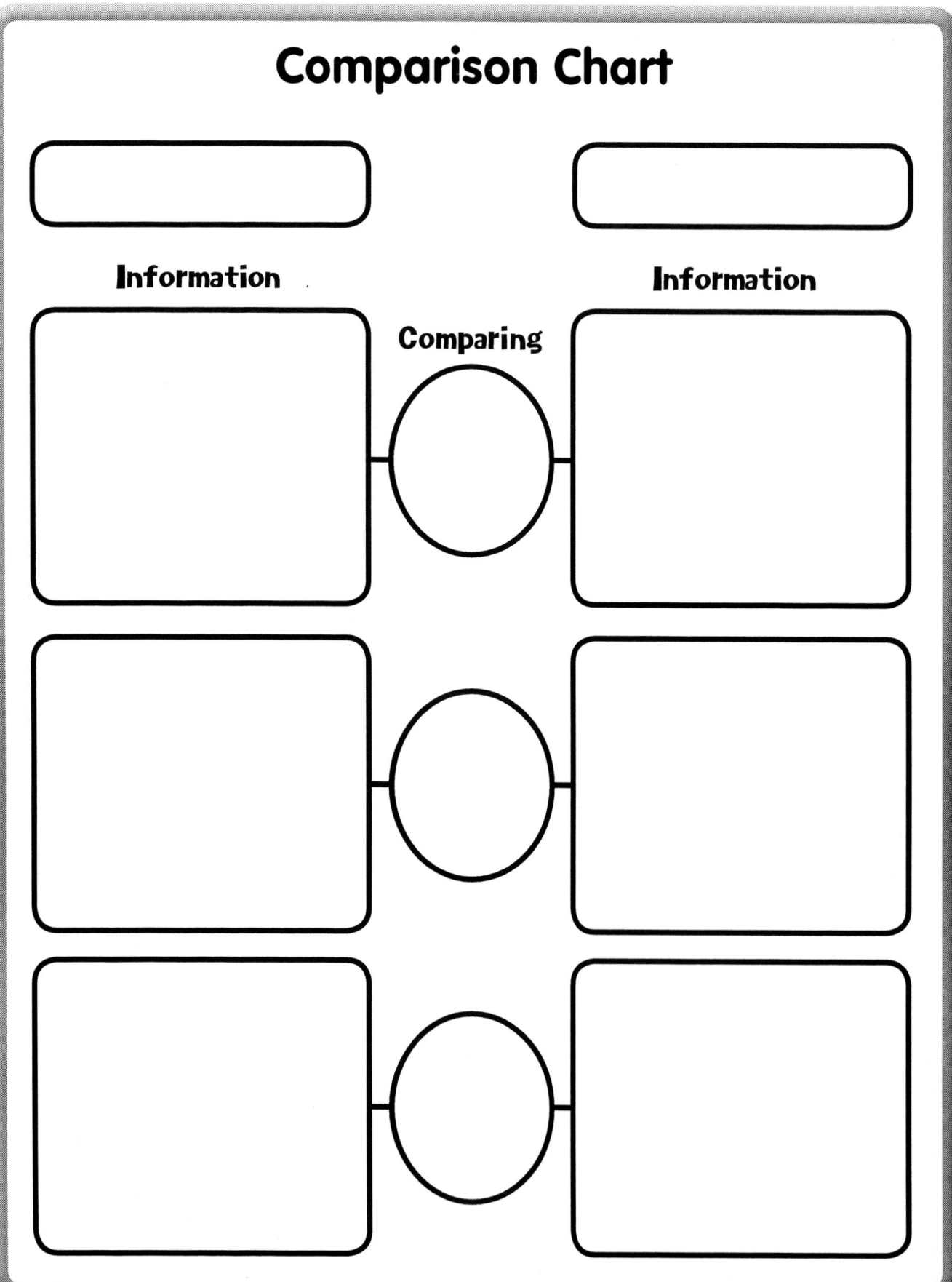

Science Expert!

_____

You are unbelievable!

Great Work!

_____

Keep up the effort!

Science Rubric

	Level 1 Below Expectations	Level 2 Approaches Expectations	Level 3 Meets Expectations	Level 4 Exceeds Expectations
Knowledge of Science Concepts	• Displays little understanding of concepts. • Rarely gives complete explanations. • Intensive teacher support is needed.	• Displays a satisfactory understanding of most concepts. • Sometimes gives appropriate, but incomplete explanations. • Teacher support is sometimes needed.	• Displays a considerable understanding of most concepts. • Usually gives complete or nearly complete explanations. • Infrequent teacher support is needed.	• Displays a thorough understanding of all or almost all concepts. • Consistently gives appropriate and complete explanations independently. • No teacher support is needed.
Application of Science Concepts	• Relates science concepts to outside world with extensive teacher prompts. • Application of concepts rarely appropriate and accurate.	• Relates science concepts to outside world with some teacher prompts. • Application of concepts sometimes appropriate and accurate.	• Relates science concepts to outside world with few teacher prompts. • Application of concepts usually appropriate and accurate.	• Relates science concepts to outside world independently. • Application of concepts almost always appropriate and accurate.
Written Communication of Ideas	• Expresses ideas with limited critical thinking skills. • Few ideas are well organized and effective.	• Expresses ideas with some critical thinking skills. • Some ideas are well organized and effective.	• Expresses ideas with considerable critical thinking skills. • Most ideas are well organized and effective.	• Expresses ideas with in-depth critical thinking skills. • Ideas are well organized and effective.
Oral Communication of Ideas	• Rarely uses correct science terminology when discussing science concepts.	• Sometimes uses correct science terminology when discussing science concepts.	• Usually uses correct science terminology when discussing science concepts.	• Consistently uses correct science terminology when discussing science concepts.

Notes: _____

_____

_____

Science Focus

Student's Name	Knowledge of Science Concepts	Application of Science Concepts	Written Communication of Ideas	Oral Communication Skills	Overall Mark

ANSWER KEY

Unit: Diversity of Living Things

Living Things, pages 2–3

1. Sample answer: A rabbit is a living thing because it is made up of many cells. It drinks water and eats things such as plants and grasses. It gets energy from eating and drinking so it can grow from a baby rabbit to an adult rabbit. It breathes air so it can get oxygen. It reacts to its environment by running away from animals that might eat it. Rabbits can reproduce.
2. gold ring: rock; cheese: milk from a cow or goat; paper: tree; wool scarf: sheep; glass jar: silica; No items are living.
3. Sample answer: Robins dig in the ground for earthworms and insects. They eat berries off bushes. They drink water from puddles.
4. Temperature is part of habitat. A habitat provides an organism with things it needs to survive. A suitable temperature is one of those things.

Classifying Organisms, pages 4–5

1. Sample answer: Scientists discovered more about organisms. What they discovered may not have fit with the earlier system, so they changed it.
2. If scientists used different classification systems, it would be difficult to share information.
3. Sample answer:

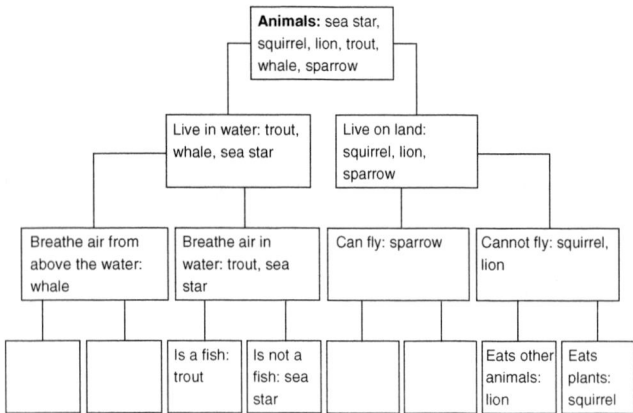

The Animal Kingdom, pages 6–7

1. Because it has the same characteristics as all animals: it is multicellular, eats other organisms such as flies, moves by itself, and lays eggs to reproduce.
2. Sample answer: Vertebrates are easier to see.
3. Because they all have jointed legs and exoskeletons.
4. wings

5. Sample answer:

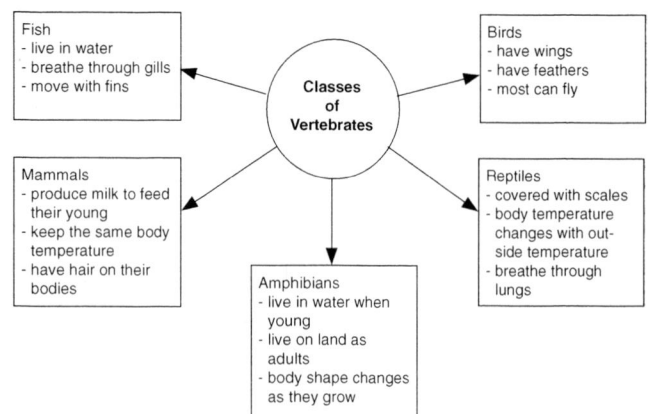

The Forgotten Kingdoms, pages 8–9

1. Sample answer:

	Cell structure	Getting nutrition	Examples of organisms	Habitat	What they do or how they affect us
Monera	- unicellular - most basic cell structure	- absorb nutrients from outside their bodies	- bacteria	- everywhere - in air, soil, water, in extreme temperatures, and in harsh chemical environments	- turn milk into cheese and yogurt - keep us healthy - make us sick - decompose waste - help make medicines
Protista	- mostly unicellular - more complex cell structure than monerans	- some make their own food - some absorb nutrients - some consume other organisms	- algae - amoebas - slime molds	- mostly in water - in damp soil - in animal bodies	- cause diarrhea and stomach upset - one causes malaria - algae are at the bottom of the food pyramid
Fungi	- mostly	- feed on dead plants and animals	- mushrooms - toadstools - mildews - molds - truffles - yeast	- most live on land in dark, moist places	- produce antibiotics - we eat some types - cause diseases in people and plants - used for baking

Cells: The Building Blocks of All Organisms, pages 10–11

1. Both are alive and carry out all of the functions of life: take in nutrients, make energy, get rid of wastes, do tasks, and reproduce.
2. Answers will vary.

ANSWER KEY

Biodiversity, pages 12–13

1. Both are classified according to characteristics and start with large groups that are divided into smaller and smaller groups.
2. Ensure that students' brochures include the worldwide locations, climate, plant and animal life, an example of a food chain, and interesting facts about their chosen biome.
3. Sketches might include a variety of deciduous trees, small bushes, and wildflowers. Animals might include deer, raccoons, foxes, and birds. Non-living features include water, air, and rocks.
4. Both communities contain living and non-living things. The living things interact with each other and help each other. The living things depend on other living things for food. They also depend on non-living things for shelter, tools, and other things they need.
5. Sample answers: Humans get products such as wood to build homes and furniture, and a variety of foods that can survive a range of conditions.

Species Interactions, pages 14–15

1. Corn is on the bottom because it is a plant. The mouse is next because it eats the corn. The snake is next because it eats the mouse. The hawk is at the top because it eats the snake.
2. a) Commensalism. The weed gets to spread its seeds and reproduce. The organism that carries the burrs is not affected.
   b) Mutualism. The bacteria get food. The human gets to digest food with the bacteria's help.
   c) Parasitism. The deer tick gets food, and the animal gets bitten and loses blood.
3. Sample answer: New species might kill off existing species. Organisms that ate that species would suffer and might die off too. The invasive species might have no predators in the new area, so the population may increase rapidly. There might be so many of the invasive species that the habitat is unable to support them. Then the habitat will change and this will affect all the other organisms.

Adaptations, pages 16–17

1. Sample answer: Dandelions have a long root that goes straight into the ground. This makes it difficult to pull them out. It can regrow from any root pieces left behind. Porcupines have strong claws that help them dig for food.
2. a) Sheltered from hot weather.
   b) Can use its front teeth underwater without swallowing water.
   c) Can see above the water when most of its body is hidden underwater.
3. Drawings might address adaptations related to gathering/hunting food, movement, and self-defence.

How Biodiversity Benefits Humans, pages 18–19

1. Sample categories: Food from Plants, Food from Animals, Clothing, and Medicines.
2. It is easier to take care of one type of crop because needs are consistent. It is also easier to harvest one type of crop that matures at the same time and is the same size.
3. The lawn will survive both dry and wet conditions.

What Do You Think? page 20

Sample answers:

1. Plant flowers that attract bees or hummingbirds; install a bird bath.
2. Make displays about different habitats to educate others about their importance; plant native plants in the schoolyard.
3. Encourage people to plant a variety of plants that attract and support a variety of native wildlife; restore and save local habitats.
4. Plant trees along roads; create more green spaces; prevent the destruction of habitats to build roads or houses; make sure there are green spaces when new developments are built.

Biodiversity Crossword Puzzle, page 21

Across: 2. vertebrates 7. invertebrates 10. behavior
Down: 1. organism 3. adaptations 4. living 5. cell 6. biodiversity 8. kingdom 9. air
Sentences: Discourage students from writing sentences that incorporate the clues. For example, organism: A plant, a squirrel, and an amoeba are all examples of an organism.

Unit: Properties of Air and Characteristics of Flight

All About Air, pages 22–23

1. Sample answer: Air can move and can move other things, as with wind. Air has a temperature, you can feel a warm breeze. Hot air rises; upstairs is warmer on a hot day. Air can hold moisture, as on a humid day.
2. a) It warms the air in the bottle.
   b) The air in the bottle expanded and moved into the balloon. I can see that the balloon inflated.
   c) Air expands when heated.
   d) The balloon will deflate because the air will cool and take up less space.

Chalkboard Publishing © 2012

ANSWER KEY

Taking Flight, pages 24–25

1. a) lift, take off
   b) more, drag
   c) weight, land
   d) Thrust, less
2. a) lift
   b) weight
   c) thrust
   d) drag

A Short History of Flight, pages 26–27

1. Sample answer:

When	Who	Accomplishment
250	People in China	- launched unmanned hot air-balloons as military signals
1500	Leonardo da Vinci	- drew airplane-like machines
1738	Daniel Bernoulli	- stated Bernoulli's principle, which explains how wings create lift
1783	Joseph and Étienne Montgolfier	- launched the first hot-air balloon with passengers
1849	Sir George Cayley	- flew the first glider with a person on board
1896	Otto Lilienthal	- made about 2000 glider flights
1903	Orville and Wilbur Wright	- first to fly in a controlled and powered aircraft

Flying Through the Sky, pages 28–29

1. The flap changes the shape of the wing, so there is more drag.
2. Flat, so there will be less drag.
3. Flaps are tilted up on one wing and down on the other so that wind pushes each wing in opposite directions.
4. A bird could push the back of its wings down to increase drag and slow itself down.
5. thrust

Experiment: Build an Aircraft, page 30

Findings will vary. Look for evidence that students have applied and/or assessed the forces of flight on their aircraft.

Aircraft with Motors, page 31

1. Sample answer: A helicopter does not need a runway. A helicopter can also hover and maneuver more precisely.

Aircraft Without Motors, page 32

1. Sample answer: Direction is determined by the wind, they do not move as fast, and carry very few passengers.
2. They have no engine, and hang-gliders are made with shapes and materials that are similar to kites.

Space Flight, pages 33–34

1. a) engine
   b) air pushing up on the fuselage (minimally)
   c) air passing the fuselage
   d) Earth
2. Sample answer: Carrying enough food, water, and oxygen, waste disposal, effects of microgravity and radiation, boredom, and dealing with serious medical problems during the trip.
3. Sample answer: Satellites cost a lot to build, launch, and maintain. Eventually they may become dangerous when they are just obstacles in orbit and when they fall back to Earth.

Animal Fliers, page 35

1. A lighter animal has less weight to overcome when they fly.
2. Insects: can turn their wings; Bats: flap their spread-out fingers to fly; Birds: flap their wings to propel them through the air.

Furred and Finned Fliers, page 36

1. Sample answer: light materials, large wings, powerful launch device, flaps for steering

Airplanes, pages 37–38

Sample answers.

1. Pros:
   - fast long-distance travel for work, vacation, and to see family
   - fast delivery of parcels and medical supplies
   - small airplanes are useful to some farmers
   - people can visit and learn about other countries and cultures
   - fly sick or injured people in remote places to a hospital
   - a rewarding hobby
   - help put out forest fires
   - search and rescue

   Cons:
   - air and noise pollution that causes stress and other health problems
   - germs spread quickly on planes
   - animals can be scared or stressed by plane noise and vibrations
   - vibrations from planes can damage buildings
   - crop dusting from planes can cause chemical pollution
   - chemicals used to de-ice planes pollute waterways
   - security and flight delays at airports cause stress and other health problems
   - planes use up lots of fossil fuel, which is a non-renewable resource
   - few people survive a plane crash

ANSWER KEY

- producing the materials used to build airplanes can harm the environment

2. a) They like airplane travel because their jobs depend on it.
   b) They dislike it because of the noise pollution and the health problems.
   c) They appreciate it because it allows them to spend less time traveling, and have access to distant medical services.

Flight Quiz, pages 39–40

1. lift, weight/gravity, drag, thrust
2. Air takes up space.
3. Air can be compressed.
4. Air has mass or weight.
5. a) Air that is moving quickly exerts less pressure than air that is moving more slowly. The air that is moving under the wing is moving more slowly than the air moving over the wing, so the air under the wing exerts more pressure.
   b) lift
6. slows it down
7. engines
8. Controlling the balance of drag on different parts of the plane helps the pilot maneuver.

Unit: Electricity

Using Electricity, pages 41–42

1. 1–air conditioner, 2–refrigerator, 3–lights, 4–clothes dryer, 5–television, 6–microwave
2. Sample answer: street and traffic lights, signs, home lighting, electric garden tools, bicycle light
3. Sample answer: cell phone, laptop computer, flashlight, portable music player, watch, smoke alarm
4. Sample answer: We would have to cook on the camp stove and use candles at night. We would not be able to use the radio, television, or computers. No hot water would come from the tap.

Current and Static Electricity, page 43

1. flashlight, power lines in a house, solar calculator, light switch
2. a) opposite
   b) same
3. Clothes rub together in the dryer, creating friction and building up charge. Pieces of clothing with opposite charges will stick together.

Experiment: Electric Cereal, pages 44–45

1. Expected results:
   5. a) When the comb got close to the cereal, the cereal swung to touch the comb and stuck to it.
   5. b) The cereal swung away from the comb.
   6. As the comb got closer to the cereal, the cereal moved away from the comb.
2. Friction creates a charge on the comb when it is rubbed against the wool.
3. This is static electricity since the energy does not flow along a path.
4. Test whether a comb attracts the cereal if it is not rubbed on wool, or test a different type of cereal.

Brain Stretch, page 46

The gloves would need to have some type of conductive material for you to work the touch screen. Rubber, plastic (e.g., fleece), or wool gloves would not work.

Conductors and Insulators, pages 46–47

1. Yes, because there is exposed wire and electricity can flow out of it and through the puddle to reach you.
2. Rubber handles prevent any accidentally touched electricity from flowing through the tool into the person.
3. Electricity could escape through the frayed cord, electrocuting someone.
4. Conductors: coin, stove element, safety pin, magnet, touch screen, earphone wire; Insulators: pencil, plastic button, rubber boots, glass, tires, wood table

Using Water to Produce Electricity, pages 48–49

1. a) Unhappy since their land will be flooded.
   b) May be happy for a new lake to use, or unhappy at the loss of the river.
   c) Happy for the new jobs during construction.
   d) Unhappy because of the threat of flooding if an earthquake damages the dam.
2. Answers will vary. Look for supported arguments.

Using Wind to Produce Electricity, pages 50–51

1. The lighter the blades, the less wind is needed to make them turn. A wind turbine with very light blades will be able to produce electricity when the wind is not very strong. Heavier blades require stronger wind to make them turn.

ANSWER KEY

2. Mining for metal and the rocks required for concrete can pollute the environment and spoil the natural beauty of the landscape. Changing ore from the mine into metal parts for the wind turbine creates pollution. Energy is required to produce the metal parts and the concrete, and this energy may be produced in ways that create pollution.

3.

Similarities	Differences
- produce electricity - use natural, renewable sources of energy - have a turbine with blades - use a generator with moving magnets and copper coils - produce electricity	- hydroelectric power plants use the energy of moving water; wind turbines use the energy of moving air - hydroelectric power plants must be built near water; wind turbines must be in windy areas that are flat and open - wind turbines can be dangerous for flying animals; hydroelectric power plants create hazards for people and barriers to wildlife such as fish - hydro dams flood large areas of land and change the flow of the river; wind turbines do not change the landforms or waterways

Transforming Energy, page 52
1. a) sound and light
   b) heat
   c) motion and sound
   d) – f) Sample answers: computer—sound and light; flashlight—light; space heater—heat; power drill—motion

All About Electrical Circuits, pages 53–55
1. a) Parallel, because there is more than one path through which the electric current can flow.
   b) Yes, because the electric current can flow past the burnt out bulb to bulb B.
2. The switch creates a gap in the circuit.
3. The rest of the bulbs will stay lit and you can see which one is burnt out.
4. a) It must be a conductor because electricity is able to get to the light bulb to make it light up.
   b) Metal, water (ice), or a person

Life Without Electricity, pages 56–58
Sample answers: More efficient appliances can save in all cases.
1. a) Today: electric stove, microwave oven, toaster, kettle; Save: use another heat source such as gas
   b) Today: refrigerator and freezer; Save: smaller appliances, cooler with ice, fill the freezer so it works more efficiently, do not hold the fridge door open
   c) Today: light bulbs; Save: turn off lights when not in use, use candles or gas lights
   d) Today: water heater sends warm water to taps; Save: solar water heater, turn down the temperature, use less water, use an on-demand water heater
   e) Today: washing machines, dryers, and electric irons; Save: wash and dry only full loads, hang to dry, cold water
   f) Today: phone, emails, websites, and text messages; Save: meet in person

Electricity Crossword, page 59
Down: 1. current **2.** plastic **3.** conductor **5.** load **6.** charge **8.** glass **9.** battery
Across: 2. power **4.** insulator **7.** static **8.** generator **10.** parallel **11.** zero

Unit: Space

Our Sun Is a Star, page 62
1. The Sun's heat warms the earth, powers the rain cycle, and creates wind. Sunlight makes plants grow, which provides us with food and oxygen.

Brain Stretch, page 62
Posters will vary. Make sure students' poster includes diagrams and labels.

Experiment: Create a Sunset, page 63
1. Light beam, maybe making some colors on the paper.
2. Reddish colors. Explanation: At sunset, the sunlight passes through much more atmosphere, so most of the blue in the sunlight is scattered. That leaves the reds and oranges to color the sunset.

ANSWER KEY

Planets in Our Solar System, pages 64–66

1. 3
2. Mars has the highest-known mountain.
3. Venus is the hottest planet.
4. Mercury is the smallest planet.
5. Jupiter is the largest planet.
6. Earth is the densest.
7. Mercury has the greatest daily temperature change.
8. Uranus spins on its side.
9. a) Mercury **iii)** peppercorn
   b) Venus **v)** black bean or **vii)** pea
   c) Earth **v)** black bean or **vii)** pea
   d) Mars **vi)** kernel of corn
   e) Jupiter **ii)** gum ball or **iv)** large marble
   f) Saturn **ii)** gum ball or **iv)** large marble
   g) Uranus **i)** small number cube or **viii)** small marble
   h) Neptune **i)** small number cube or **viii)** small marble

Moons, pages 67–68

1. Diagrams should show that the Moon is in the darkest part of Earth's shadow.
2. a) The Moon phases are already in order but students may start their observation at any point along the set, affecting where #1 starts.
   b) From Earth, the angle at which we observe the Moon's sunlit side changes, meaning we see less or more of the Moon.

They Came from Outer Space, page 69

1. From largest to smallest: asteroids, meteoroids, comets.
2. Sample answer: Perseids are a group of meteors that we see streaking across the sky (like a shower) in the area where the Perseid constellation is found.

Experiment: Mass of Asteroids, page 70

1. Answers will vary.
2. metal was heavier
3. Metal asteroids have more mass than stone asteroids of the same size do.
4. A comet contains a lot of frozen material, so it would be lighter/less dense.

Explore the Night Sky, page 71

1. a) – b) Diagrams should have a detailed title and show several starry objects.
   c) Sample answer: We looked in different directions or at different days/times.
2. a) – b) Answers will vary. Encourage students to research a culture they relate to.

Astronauts and Space Travel, pages 72–73

1. Sample answer: Food, air, and water because there is none in space; computer with photos of friends, music player, e-book reader, and Internet access for entertainment, keeping and analyzing records, and communication; therapy bands for resistance exercise; sleeping bag because blankets would not stay in place while sleeping; ear plugs to block out others' sounds; and tools to fix things.
2. Sample answer: protection from the airless, cold, radiation-filled outer space; storage for supplies (including food) and waste (dirty clothes, human waste, food packaging, etc.); exercise area; sleeping and bathroom spaces; place and equipment to perform experiments; communication with Earth; windows.

Other Spacecraft, pages 74–75

1. Sample answer: Orbiters have no danger of crash-landing and no need to launch to leave the planet. They do not need to withstand conditions or terrain on the planet. Landers can take samples and examine the planet on a microscopic level.
2. Sample answer: Rovers have to withstand fast acceleration, high and low temperatures, pressure, and dust, and must function without repair. They must be small enough and light enough to fit on a spacecraft and there has to be a way to remove them from the spacecraft when they reach their destination. The rover should be remotely controlled, move over the landscape, and send pictures.
3. A rover can move to get more or better samples or to take pictures from different angles.

Astronaut Research Report, page 76

Have students share their reports with the class.

Space Technology, pages 77–79

1.–2. Answers will vary. Look for reasoned arguments.
3. Similarities include the confined space and need for life support.
4. a) Fresh **vi)** Food is not preserved so it must be eaten quickly.
   b) Intermediate Moisture **ii)** Food with some water removed; water is not replaced before eating.
   c) Irradiated **iii)** Food is cooked and packed in foil pouches. It is sterilized by radiation so it can be kept at room temperature.
   d) Natural **v)** Food is ready to eat and stored in flexible pouches.
   e) Rehydratable **i)** Food is freeze-dried to remove water. Water is replaced before eating.
   f) Thermostabilized **iv)** Food is heated to kill bacteria so it can be stored at room temperature.
5. Answers will vary.

ANSWER KEY

Animals in Space—Fair or Not? page 80
1. Any reasoned answer is acceptable.

Those Puzzling Planets, pages 81–82
Across: 2. moon **4.** orbit **5.** Mars **7.** Saturn **10.** Earth **11.** Neptune **13.** Uranus **14.** Mercury **17.** Sun **18.** asteroids **Down: 1.** solar system **3.** constellation **6.** comet **8.** meteor **9.** Venus **12.** Jupiter **15.** Canadarm **16.** galaxy